the undersea network

SST sign, storage, transmission

A series edited by Jonathan Sterne and Lisa Gitelman

NICOLE STAROSIELSKI

the undersea network

Duke University Press Durham and London 2015

Printed in the United States of America on acid-free paper ∞
Designed by Courtney Leigh Baker
Typeset in Gill Sans and Whitman by BW&A Books, Inc.

Library of Congress Cataloging-in-Publication Data
Starosielski, Nicole, 1984–
The undersea network / Nicole Starosielski.
pages cm—(Sign, storage, transmission)
Includes bibliographical references and index.
ISBN 978-0-8223-5740-7 (hardcover : alk. paper)
ISBN 978-0-8223-5755-1 (pbk. : alk. paper)
ISBN 978-0-8223-7622-4 (e-book)
1. Cables, Submarine. 2. Fiber optic cables. 3. Telecommunications lines. 4. Telecommunication cables. 5. Communication, International. I. Title. II. Series: Sign, storage, transmission.
TK5103.15.S737 2015
384.3'2—dc23 2014037968

Cover art: Cable landing, Guam. Photo by the author.
Concept map illustrations for each chapter drawn by Cameron Rains.

Duke University Press gratefully acknowledges the support of the Humanities Initiative Grants-in-Aid, New York University, which provided funds toward the publication of this book.

for my family

CONTENTS

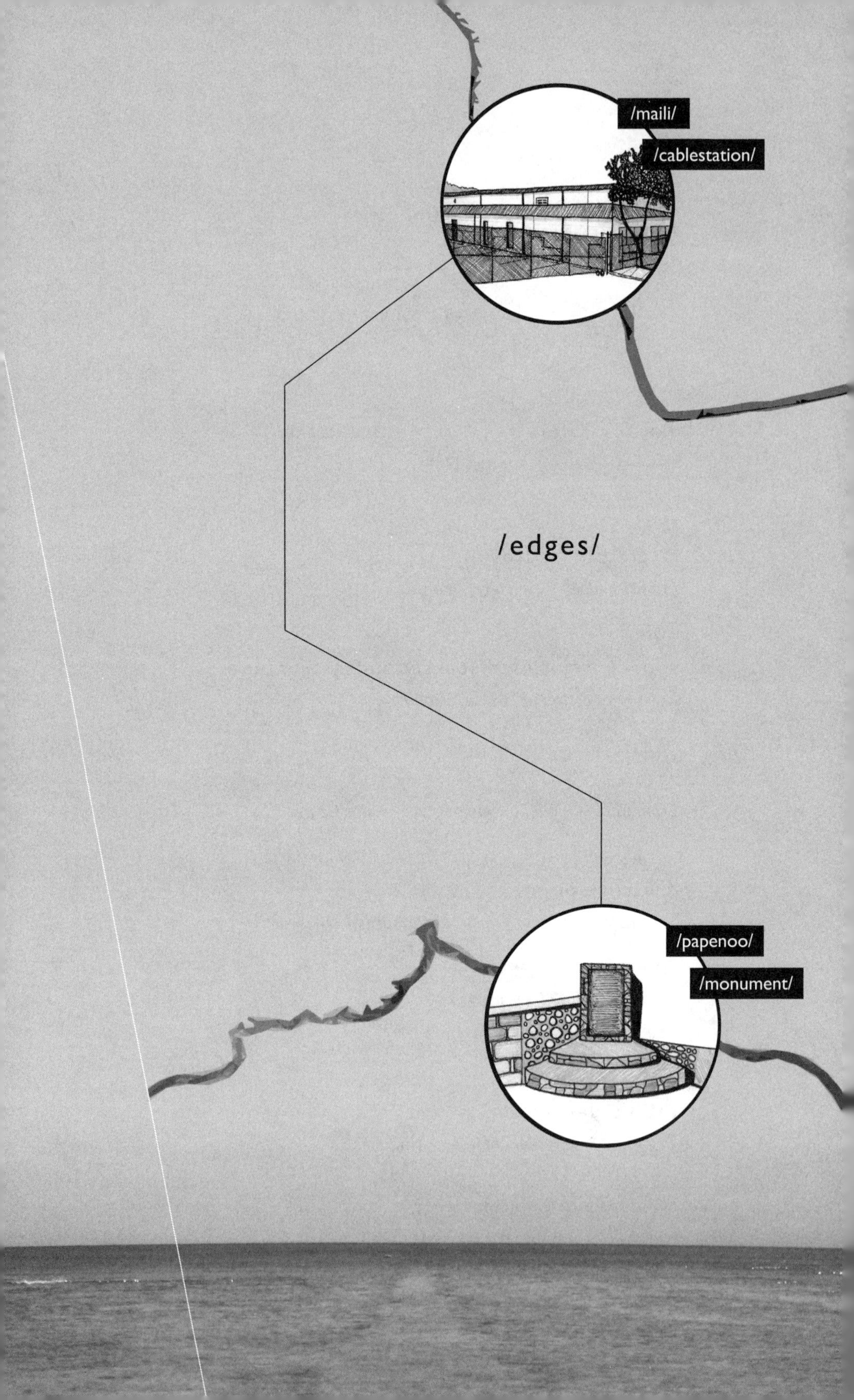
/maili/
/cablestation/
/edges/
/papenoo/
/monument/

PREFACE

edges

O'ahu, Hawai'i, United States

I am standing on Electric Beach, on O'ahu's west shore—a beach named for the large power plant towering behind it and known for regular car burglaries. Three men are casually fishing off the edge of the point. Families are having barbeques. Posing as a tourist with a camera, I crouch down to take pictures of a manhole covered in rust-colored dirt (figure P.1). Underneath the manhole, a fiber-optic cable surfaces, bringing information encoded in light waves from the other Hawaiian islands. Within thirty miles of this point, cable systems extend directly to California, Oregon, Fiji, and Guam and reach onward to Australia, Japan, and much of East Asia. Though they are at the edge of the United States and the periphery of Americans' vision, O'ahu's cable landings establish Hawai'i as a critical node in our global telecommunications networks. Manholes, such as the one beneath my feet, are some of the few sites where cable systems appear in public space. It is by looking down, rather than up to the sky, that we can best see today's network infrastructure.

FIGURE P.1. O'ahu cable landing.

Following the cable route toward the ocean, I find a path carved out by foot traffic, demarcating a connection between land and sea. As I turn from the water's edge to head back to the cable station, I experience a moment of disconnect. In front of me is a small cave containing hanging clothes, stockpiled chairs, and collected water. The same beach that makes possible the landing of communications cables—infrastructures that accelerate the movement of information across oceans—is also a temporary dwelling for some of the least mobile Hawaiians. When I travel up O'ahu's west shore, I meet residents who help me to make sense of the apparent contradiction. The histories of cable laying, militarization, and economic deprivation in the area are intertwined: the modes of spatial organization that have enabled O'ahu to become a communications hub have also displaced local residents to tent cities on coastal beaches. In reaction, residents have developed a territorial politics that challenges the cable companies' extensions through the shore. This is not the only place where local conflicts over territory obstruct network development. Across the Pacific, companies have to apply for extensive permits to traverse the cable landing point, and at times their projects are diverted to alternate routes or stopped altogether.

In the early 2000s, toward the end of the fiber-optic cable boom—a period of intense infrastructure building that coincided with the emergence of the Internet—Tyco Telecommunications built a station in the town of Ma'ili, several miles north of Electric Beach. Although cable stations were once central workplaces that enriched and enlivened communities, today they are more often inaccessible buildings that bring little visible benefit to the surrounding area. This link to the information highway, shuttling signals between Asia and the United States, had neither on-ramps nor off-ramps as it extended under the houseless people of the west shore. The station is located in the center of the town, next to an elementary school. Children play nearby, their voices filtering across the lawn, over heaps of trash in an adjacent lot, and through the station. When I visit the site, I am not surprised to see that it is abandoned and has bullet holes in its windows. Tyco Telecommunications encountered community resistance when it decided to lay cable beneath the town, and it could not afford to bring the cable ashore in Hawai'i because of the eventual bust of network development. The telecommunications worker who brought me here speculates that the station's heightened visibility and its proximity to the school intensified Tyco's difficulties.

This visit to O'ahu in 2009 was my introduction to the geography of undersea networks, and it remains a formative memory as I write this book. In O'ahu I first recognized the resolute materiality of network infrastructure and its entanglements with the turbulent histories of the Pacific, ranging from local cultural practices to large-scale projects of colonization and militarization. This propelled my journey across the Pacific to track the telegraph, telephone, and fiber-optic cable routes from North America (California, Oregon, Vancouver, and Washington) through islands that have been critical to transpacific networking (Fiji, Guam, New Zealand, Tahiti, and Yap) to economic centers across the ocean (in Australia, Hong Kong, Japan, the Philippines, and Singapore). At these sites, I traced the institutional histories of cable networks, documented their technological installations, and chronicled the range of cultural uses for cabled spaces. Cable routes are not only makeshift homes but also places for dumping trash, areas to be preserved and protected, sites for recreation, and even sacred grounds. Digging into the histories of our fiber-optic systems, I found that the currents of Internet traffic—often seen as flattening the Pacific Rim—have instead gained traction in its diverse cultural environments.

Papenoo, Tahiti

Two years later and a thousand miles south, I pull up to a second school on the island of Tahiti. By this time, my visits to cable stations have become routine. I look for an unmarked and nondescript industrial building with few windows, surrounded by surveillance cameras and guarded by barbed wire.

Contemporary cable systems are critical to the functioning of our global information sphere—they transmit almost 100 percent of intercontinental Internet traffic—and are embedded in a landscape of security. Failing to find the station, I park on a dirt strip outside a school and wander into its open courtyard. Students pull at two ends of a long rope in a game of tug-of-war (figure P.2). I approach the woman watching these children and ask her about Honotua, Tahiti's first fiber-optic cable, which was laid earlier that year. She calls a young boy over from the yard. "Le câble!" she points. He runs toward the ocean and I follow, taking snapshots as I duck through the buildings. Arriving at the back of the school, I encounter a sight as striking to me as the rudimentary inhabitation of O'ahu's cable landing. Here stands a stone monument, about five feet tall. A large black plaque is mounted on its face. An inscription in Tahitian, English, and French reads:

> In memory of the people of Papenoo and of Hawai'i, who established ties in the past:
> Tapuhe'euanu'u from Tapahi, who, fishing from his canoe, caught Hawai'i the Great,
> Te'ura-vahine from Ha'apaiano'o, the goddess Pere, who sought refuge in the volcano of Hawai'i the Great,
> Mo'iteha, King of Hawai'i, who came back to Tahiti to build his marae Ra'iteha at Mou'a'uranuiatea,
> Ra'amaitahiti, his son, King of Tapahi, who brought his drum to Kaua'i,
> To revive these ancient connections, Honotua was made: The submarine cable that links Tahiti to Hawai'i.
> After quietly undulating in the deep sea, it has landed here, at Mamu (silence).
> Hopefully human ignorance will dissolve into silence and only knowledge will be conveyed.

This is the only landing point at which I have ever seen an active cable memorialized. Instead of being hidden, with only a manhole to indicate its location, Honotua is marked proudly for anyone to see. The plaque does not describe undersea cables as a new technology but instead highlights the continuity between the light waves that transmit information and the ocean waves that have carried islanders across the Pacific. Although in Hawai'i undersea cables were resisted by residents, who perceived them as part of a colonial legacy, in Tahiti cable infrastructure is displayed as an important site in local education, integral to the transmission of cultural knowledge. Regardless of what

FIGURE P.2. A game of tug-of-war, Papenoo, Tahiti.

purpose the cable will actually be used for, this link is commemorated not simply because it hooks Tahiti into a global network, but also because it is seen as building on Tahitians' past cultural connections. The turbulent environments of Hawai'i, however, had created a ripple effect across the system, affecting the Tahitian cable's geography. Honotua could have landed on O'ahu, but the contentious spatial politics of the west shore meant that companies must bury cables there by drilling a conduit horizontally under the sand. This strategy of insulation keeps the cable out of view and doesn't disrupt local road traffic, but it reportedly costs over fifty times more than simply digging a trench, ultimately deterring Honotua's owners. Instead, this cable terminates on Hawai'i's big island, a new outpost for undersea networks, geographically separated from the historical concentration of transpacific systems. Cable networks not only build on past cultural connections, but they become entangled in contemporary cultural conflicts.

The link between two locations in a network, such as the connection between Hawai'i and Tahiti, is termed an "edge" in network theory, an appropriate term given that we rarely see beyond its horizon. Edges are often drawn as a simple line between two nodes, a vector that stands independent of time and place. Rather than take such connections for granted, this book moves through the environments of our undersea network, into the routing arrangements, cable stations, landing points, and subaquatic spaces in which links have been constructed. It focuses our attention on the geography of cable construction, operation, and contestation, and on the companies that are themselves caught in a tug-of-war between the need to insulate currents from their environments—via walls, beaches, or other protective measures—and to connect them with preexisting circulations of meaning and value. Exploring the materiality of such edges reveals how our undersea network, as well as the connections it enables, has been made possible only by the deliberate manipulation of technology, cultures, politics, and environments, all of which remain invisibly enfolded in the lines between nodes.

Statement on *Surfacing*

The Undersea Network weaves a set of narratives across our transoceanic cable systems, connecting rural cable stations with submarine ridges, remote islands to urban centers, and large-scale historical forces with localized conflicts. From each of these sites a network extends outward to a myriad of technologies, actors, and events. Every cable station connects to an undersea cable system, as well as to a set of culturally specific practices of operation. Each island is embedded in a broad social and political history. Even localized conflicts have been shaped by varied corporate and governmental actors. The following stories traverse only some of these vectors.

Surfacing, a digital map of undersea cables, draws readers deeper into these hidden networks. In this online system, the reader can dive into the photographic archives of individual cable routes, explore the local histories of cable stations and landing points, and navigate the numerous connections between nodes. *Surfacing* provides a nonlinear way to access our undersea network, one that is geographically rather than narratively oriented.

Surfacing connects to this book via a series of keywords—portals between print and geography—that are indicated at the beginning of each chapter in a concept map. To move to a site in *Surfacing*, simply type the keyword after "surfacing.in":

To access
/centralcalifornia/
visit:
surfacing.in/centralcalifornia

ACKNOWLEDGMENTS

Like all signals, this book bears traces of its environments.

I was inspired to start a project about media under water at the University of California, Santa Barbara (UCSB), a campus perched at the edge of the ocean. I am grateful for the generosity of the faculty, graduate students, and staff in the Department of Film and Media Studies. I could not have asked for a better dissertation chair, mentor, and collaborator than Lisa Parks, whose advice it was to first look at undersea cables, and whose insight and intellect marks this book. My dissertation committee members, Janet Walker, Constance Penley, Rita Raley, and Greg Siegel, provided critical feedback and helped me to establish the project's foundations. I am also thankful for the intellectual support of UCSB faculty members Peter Bloom, Edward Branigan, Bishnu Ghosh, Anna Everett, Dick Hebdige, Jen Holt, Stephanie LeMenager, Mireille Miller-Young, Bhaskar Sarkar, Cristina Venegas, and Chuck Wolfe. Many of my fellow graduate students took time to comment on my work in progress, and among these I would particularly like to thank Jeff Scheible, Meredith Bak, Maria Corrigan, Chris Dzialo, Sarah Harris, Anastasia Yumeko Hill, Regina Longo, Josh Neves, Dan Reynolds, and Ethan Tussey.

The project moved from California out to the Pacific in part thanks to grants from the University of California (UC) Pacific Rim Research Program, the UC Humanities Research Institute California Studies Consortium, and the UC Interdisciplinary Humanities Center. Above all, I am indebted to the numerous people who agreed to be interviewed and helped me document the histories and geographies of undersea cables. Without their interest and kindness, I would not have been able to undertake such an expansive project. Because of both the sensitive nature of the information and the small size of the cable industry, it was necessary to make many of the statements that I have quoted anonymous. Since some people's names are excluded, I have chosen to be consistent and mention none of my interviewees here, rather than highlighting only a subset of the many people who helped me with my research. However, I would like to thank those associated with the International Cable Protection Committee and SubOptic for their support—in particular, Dean Veverka, who helped me to navigate the complexities of the cable industry. Kent Bressie, Douglas Burnett, Catherine Creese, and Brett O'Riley also provided invaluable advice. In addition, I would like to thank Andy, Mel, David, Annette, Jelena, and the many other friends who accompanied me on trips to landing sites across the Pacific.

My first job out of graduate school landed me in southwest Ohio, where two years at Miami University of Ohio gave me the time to think through my field material. I am thankful to my colleagues in the Department of Communication, especially to Ron Becker, and to my department chair, Richard Campbell. The Miami University Humanities Center and Tim Melley supported me in a year-long conversation on digital networks. cris cheek and Braxton Soderman were wonderful collaborators in developing the Network Archaeology Conference and the special issue of *amodern*, in which I refined the methodology of the book.

I began this project on a beach, traveled throughout the Pacific islands, and spent an interlude in the Midwest, but I ended it in the center of a global city. At New York University, I was given time and support to finish the manuscript. For the intellectual community I found there, I would like to thank my colleagues and students, especially Lisa Gitelman, Jamie Skye Bianco, Alex Galloway, and Martin Scherzinger. The text has also been marked by my time at the University of Southern California, where Marsha Kinder, Tara McPherson, and Bill Whittington prepared me for the challenge of tackling both a book project and a related media platform. Craig Dietrich, Erik Loyer, and Shane Brennan were fantastic collaborators in developing *Surfacing*, the interactive digital map that accompanies this text. The International Cable Protection Committee and SubOptic also provided financial and logistical support to bring the final version of the project to fruition.

Chapters 4 and 5 of the book originated as the articles "'Warning, Do Not Dig': Negotiating the Visibility of Critical Infrastructures," *Journal of Visual Culture* 11, no. 1 (2012): 38–57; and "Critical Nodes, Cultural Networks: Re-Mapping Guam's Cable Infrastructure," *Amerasia* 37, no. 3 (2012): 18–27. I would like to thank Courtney Berger and the editors at Duke University Press, the three reviewers who provided key insights and vastly improved the book, and those who took the time to read sections of the manuscript and brainstorm with me about its development, including Brooke Belisle, Kelsey Brannan, Fernando Dominguez Rubio, Anitra Grisales, Shannon Mattern, and Helga Tawil-

Souri. I extend special thanks to Stefan Helmreich, Eva Hayward, and Stacy Alaimo, my fellow aquatic explorers. The illustrations that preface each chapter have been drawn by Cameron Rains, whom I am lucky to have as both a collaborator and a friend.

Finally, I dedicate this book to my family who, though they may seem outside the project, have nonetheless made it possible. To those from California: Genevieve, Josh, Lea, Lizzie, Ross, and York. And from Cincinnati: Amy, Danielle, Eva, Jen, and Judi. Finally, landing in New York: Jamie, Ben, Erica, Lily, and Matt. A special thank-you to Jeff, whose comments on things under water over the past six years permeates this book. And last, but not least, to my brother Alex, who spends his time routing power through radar systems; to my father, a photographer and a diver, who taught me how to look at the world through a lens; and to my mother, whose many years of work for Continental Airlines instilled in me a love for flight.

INTRODUCTION

against flow

Undersea fiber-optic cables are critical infrastructures that support our global network society. They transport 99 percent of all transoceanic digital communications, including phone calls, text and e-mail messages, websites, digital images and video, and even some television (cumulatively, over thirty trillion bits per second as of 2010).[1] It is submarine systems, rather than satellites, that carry most of the Internet across the oceans. Cables drive international business: they facilitate the expansion of multinational corporations, enable the outsourcing of operations, and transmit the high-speed financial transactions that connect the world's economies. Stephen Malphrus, staff director at the U.S. Federal Reserve Board, has stated that if the cable networks are disrupted, "the financial services sector does *not* 'grind to a halt,' rather it *snaps* to a halt."[2] As a result, the reliability of undersea cables has been deemed "absolutely essential" for the functioning of governments and the enforcement of national security.[3] Militaries use the cables to manage long-range weapons tests and remote battlefield operations. Undersea networks also make possible new distributions of transnational media that depend on high-capacity digital exchange, from the collaborations of production companies in the United States and New Zealand on the 2009 film *Avatar* to the global coordination of *World of Warcraft* players. At the same time, cable infrastructure enables modes of resistance that challenge dominant media formations. Messages produced by the Arab Spring and Occupy movements traveled between countries on undersea cables. If the world's 223 international undersea cable systems were to suddenly disappear, only a minuscule amount of this traffic would be backed up by satellite, and the Internet would effectively be split between continents.[4]

This book traces how today's digital circulations are trafficked underground and undersea, rather than by air. It follows signals as they move at the speed of light, traveling through winding cables the size of a garden hose. En route, they get tangled up in coastal politics at landing points, monitored and maintained at cable stations, interconnected with transportation systems and atmospheric currents, and embedded in histories of seafloor measurement. Cable infrastructures remain firmly tethered to the earth, anchored in a grid of material and cultural coordinates. *The Undersea Network* descends into these layers to reveal how such environments—from Cold War nuclear bunkers to tax-exempt suburban technology parks; from coasts inhabited by centuries-old fishing communities to the homes of snails, frogs, and endangered mountain beavers—continue to underlie, structure, and shape today's fiber-optic links. From this vantage point, apparently outside the network, one can see the hidden labor, economics, cultures, and politics that go into sustaining everyday intercontinental connections. Rather than envisioning undersea cable systems as a set of vectors that overcome space, *The Undersea Network* places our networks undersea: it locates them in this complex set of circulatory practices, charting their interconnections with a dynamic and fluid external environment.

As a result, the book offers what might be an unfamiliar view of global network infrastructure. Not only is it wired, but it is also relatively centralized—far from the early vision of the Internet as a rhizomatic and distributed network. Transoceanic currents of information have been fixed along fairly narrow routes through the specialized work of a small cable industry, which has navigated natural environments, built architectures of exchange, and generated new social and cultural practices, all to ensure our media and communications safe transit through the surrounding turbulent ecologies. Rather than a strictly urban system, cables are rural and aquatic infrastructures. Conservative and yet resilient, they have followed paths that are tried and true, often following the contours of earlier networks, layered on top of earlier telegraph and telephone cables, power systems, lines of cultural migration, and trade routes

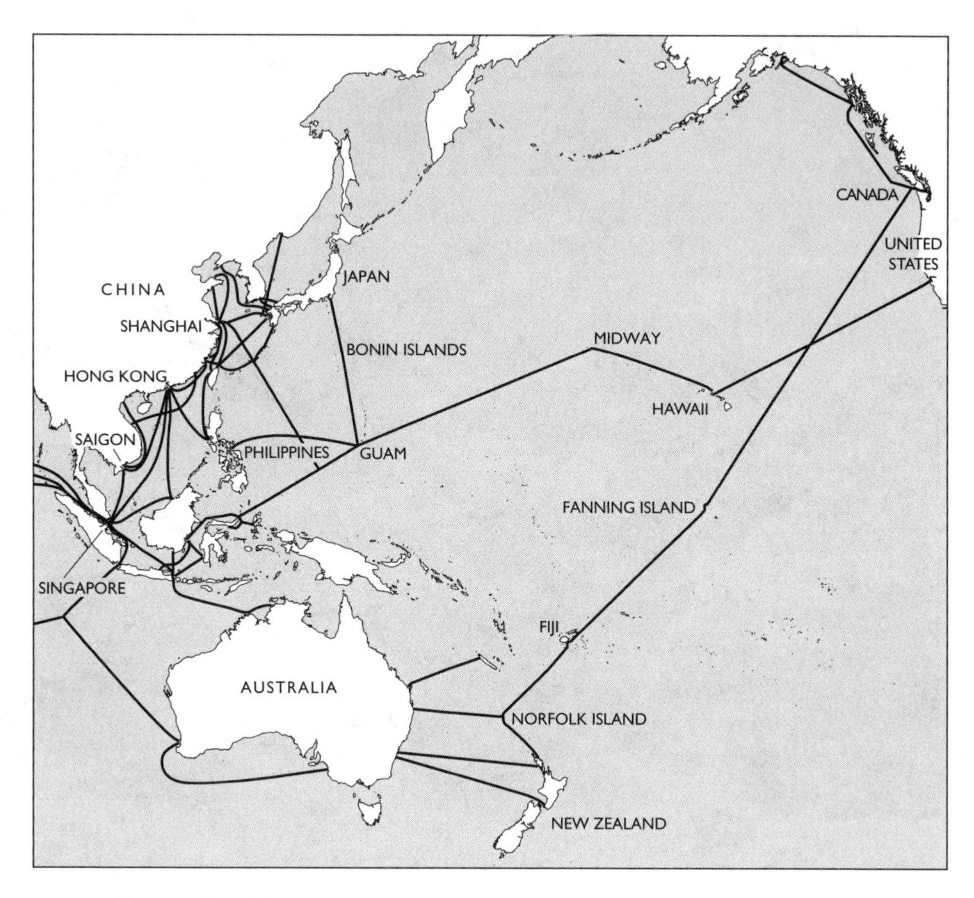

MAP I.1. Transpacific cable routes, 1922.

(figure I.1; maps I.1–I.2). All of these tend to remain outside of our networked imagination, a world defined by firmly demarcated nodes, straight and clear vectors, and graph topologies. As Alan Liu observes, a network "subtracts the need to be conscious of the geography, physicality, temporality, and underlying history of the links between nodes."[5] By bringing these geographies back into the picture, this book reintroduces such a consciousness, one might even say an environmental consciousness, to the study of digital systems.

Invisible Systems

Why have undersea cables, as the backbone of the global Internet, remained largely invisible to the publics that use them? Cable development has often been justified on the basis of cables' perceived security (as opposed to commu-

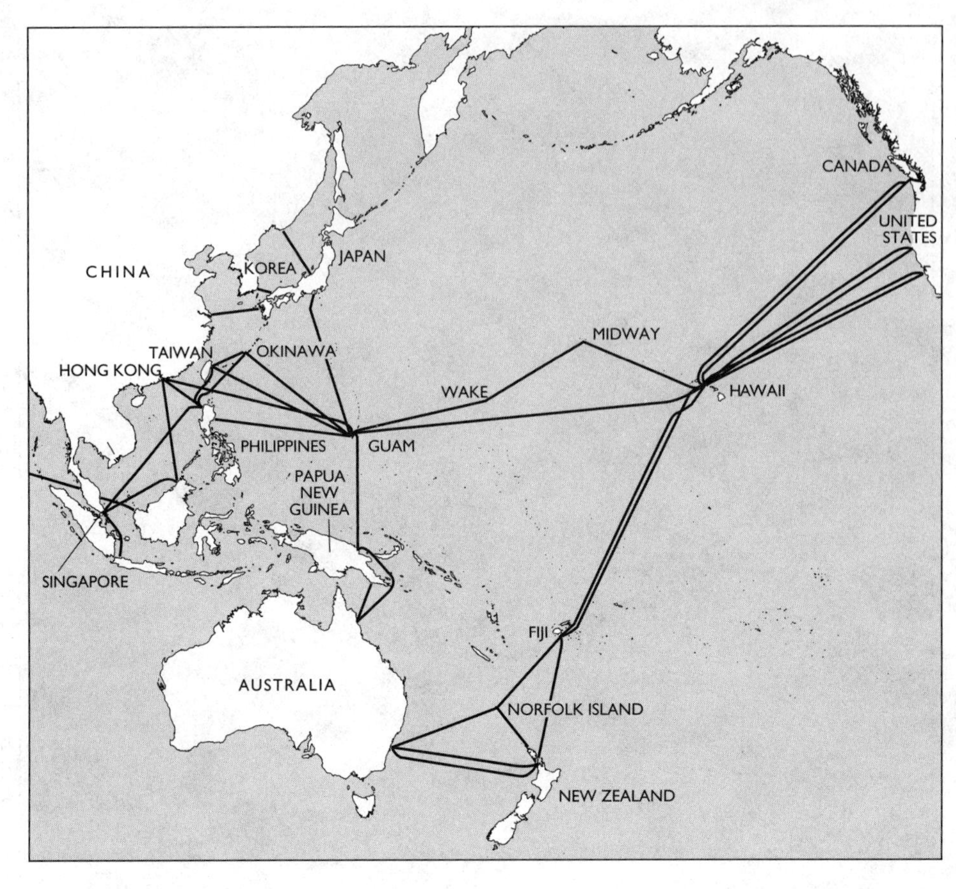

MAP I.2. Transpacific cable routes, 1982.

nications via satellite or radio, which are more easily intercepted) and information about the networks has often been withheld in a strategy of "security through obscurity."[6] After all, the reasoning goes, if the public doesn't know about the importance of undersea cables, they will not think to contest or disrupt them. The scarcity of facts circulated about cable systems also reflects the hesitance of a competitive international telecommunications industry to release information of commercial value. More than any intentional desire to obscure cable systems, their invisibility is due to a broader social tendency to overlook the distribution of modern communications in favor of the more visible processes of production and consumption. As Susan Leigh Star observes, infrastructure "is by definition invisible, part of the background of other kinds of work."[7] Many people in the cable industry perceive a general lack of public interest in their infrastructures. When I interviewed Stewart

Ash, who has worked for decades on undersea cable design and installation, he pressed me on my interest in making cables visible. "Why would you want to know?" he asked. "When you turn on a computer and you send an e-mail, do you really care how it works? No, you just want e-mail there, and you start drumming the table if it takes thirty seconds."[8]

Submerged under miles of water for decades and seemingly disassociated from our everyday lives, undersea cables are particularly difficult to connect to our imagination of media and communication. When communications infrastructures are represented, they are most often wireless: handheld devices, laptop computers, wireless routers, cell phone towers, "cloud" computing, and satellites pervade our field of view, directing our attention above rather than below and reinforcing a long-standing imagination of communication that moves us beyond our worldly limitations.[9] One cable engineer I spoke with—a manager at one of Australia's most critical cable stations—claimed that satellites are simply just "sexier" than cables. He admitted that even after his company's communications shifted to cable, they still displayed advertising suggesting that conversations were being carried by satellites, showing signals being bounced out into orbit and then back again because that was what stuck in people's minds.[10] Undersea cables, he claimed, are "not a technology that people find fascinating." Leaving the station after our interview, I observed images of satellites plastered on the side of the building.

When we do see public representations of undersea cables, these tend to divert our attention away from the materiality of the network. As I describe in chapter 2, narratives about undersea cables often focus on nonoperational infrastructure: there are films about cable planning and laying, news articles at the moments of network disruption, and histories of artifacts from obsolete systems. Cable industry publications tend to focus on capacity and feature few geographic details. The typical cable map portrays the cable as a vector that indicates connectivity between major cities or even just countries (figure I.1, maps I.1–I.2). The environments that cables are laid through—the oceans, coastal landing points, and terrestrial routes—are seen as friction-free surfaces across which force is easily exerted, and where geographic barriers are leveled by telecommunications. As Philip Steinberg has observed, this conception of space is a Western ideal that has historically been linked to the expansion of capitalism.[11] Depicting the ocean and the coasts as deterritorialized naturalizes the claims of actors that might capitalize on their connective capacity, such as cable companies, and presents an obstacle to those that might claim it as a territory, such as nations.

Fiber-optic cables have also remained largely absent in the field of media

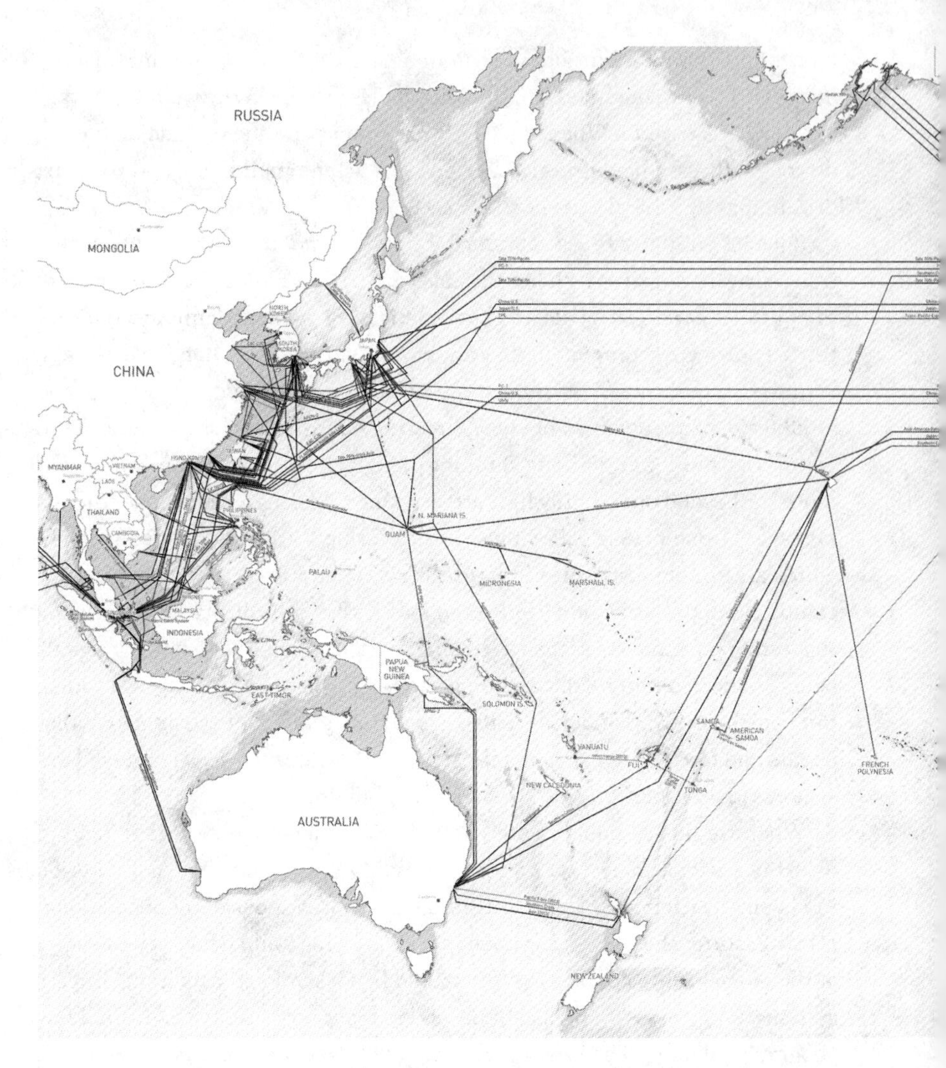

and communications studies, which has focused on the content, messages, and reception of digital media and paid less attention to the infrastructures that support its distribution. Analyses of twenty-first-century media culture have been characterized by a cultural imagination of dematerialization: immaterial information flows appear to make the environments they extend through fluid and matter less. Mark Taylor, arguing that the contemporary network economy is made possible by ever-extending dematerialization, writes that the "Internet is really nothing more than codes and protocols that enable computers to communicate."[12] When cables become an object of study, it is almost always as a form of old media. Historians of technology have carefully detailed the beginnings of telegraph cable networks in the 1850s and 1860s and the extension

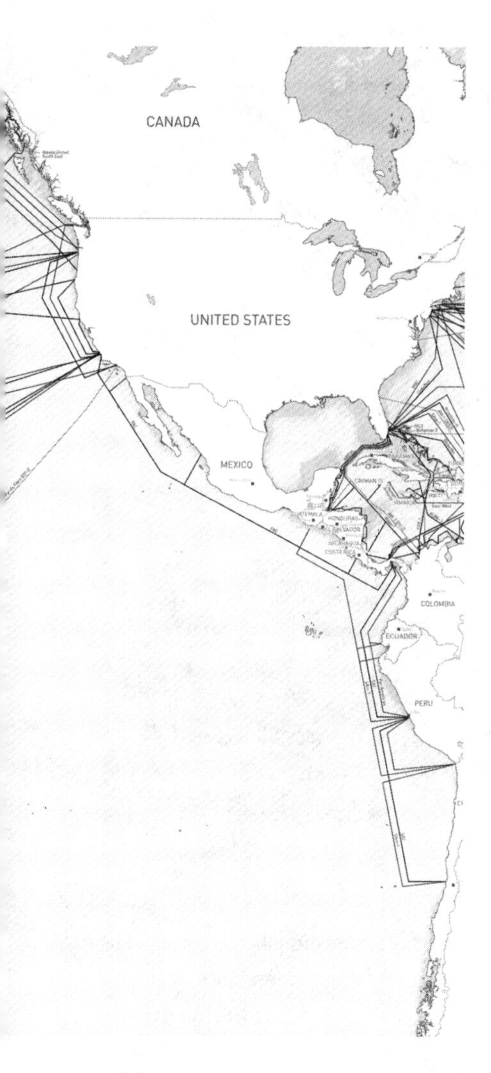

FIGURE I.1. Transpacific cable routes, 2012. Courtesy of TeleGeography, www.telegeography.com.

of these systems through the 1940s in the context of British colonial rule, conflicts between nation-states, and a global media economy.[13] There are no major studies that detail the cultural geographies of undersea coaxial cables laid between the 1950s and 1980s, the undersea fiber-optic cables of the 1990s, or the links between these newer forms and older cable systems. Cables have instead been submerged in a historiographic practice that tends to narrate a transcendence of geographic specificity, a movement from fixity to fluidity, and ultimately a transition from wires to wireless structures.

Although wired and wireless technologies are often positioned as historical competitors, cables and satellites actually have different geographic dispersions, markets, and technological affordances. Satellites, with their wide

FIGURE I.2. Advertisement for Submarine Cables Ltd., 1960s. From *Zodiac*, © Cable & Wireless Communications 2013, by kind permission of Porthcurno Telegraph Museum.

field of reception, have been more useful for rural areas and islands; they have historically been used for mass communication and have been critical in the transmission of television.[14] Undersea cables, laid on the very bottom of the ocean and surfacing only at the landing points at either end, are more efficient for point-to-point routes of dense information exchange. They also have the benefit of increased security, a consideration for military and government traffic (figure I.2). Overall, the telecommunications industry has long regarded wired and wireless forms as complementary. Achieving redundancy is critical, and the best networks have multiple routes to any single destination. Therefore, even though the percentage of signals carried by wired technologies has ebbed and flowed, they have continued to support the expansion of economic, political, and cultural networking even during the eras of radio and satellite (figure I.3).

Over the past twenty years, satellites' capacity has filled up, and conditions have shifted significantly to favor fiber-optic cables.[15] Cables are now able to

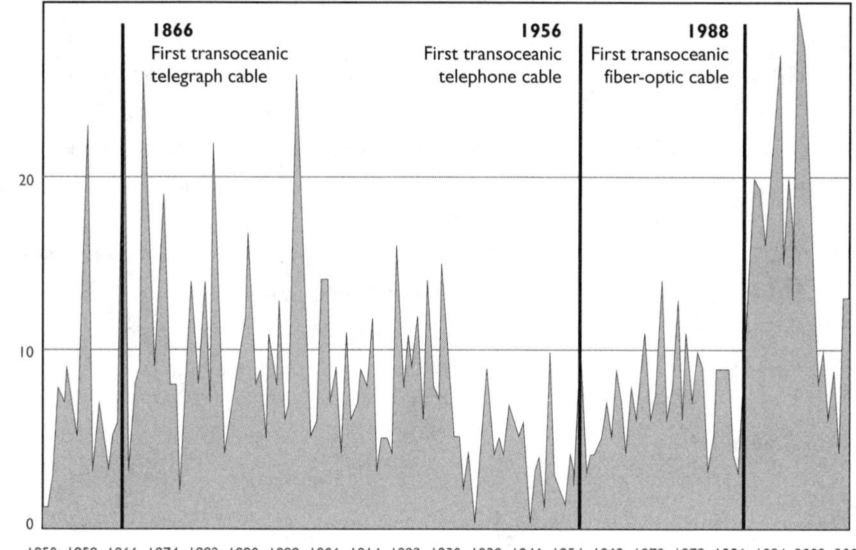

FIGURE I.3. Undersea cable systems established per year, 1850–2009. Cables remain significant even during wireless eras. Data from Burns, "History of the Atlantic Cable."

carry a greater amount of information at faster speeds and at lower cost than satellites (a signal traveling between New York and London takes about one-eighth the time to reach its destination by cable as it does by satellite).[16] With the emergence of high-definition video and high-bandwidth content on the Internet (a shift that favors cable infrastructure), the disparity between the two looks like it will only increase. Despite the rhetoric of wirelessness, we exist in a world that is more wired than ever. As Adrian Mackenzie puts it, "While the notion of wireless networks implies that there are fewer wires, it could easily be argued that actually there are more wires. Rather than wireless cities or wireless networks, it might be more accurate to speak of the re-wiring of cities through the highly reconfigurable paths of chipsets. Billions of chipsets means trillions of wires or conductors on a microscopic scale."[17] Although contemporary networking continues to depend on wired infrastructures, we lack a language—beyond terms like "a series of tubes"—to describe just how grounded these systems remain.[18]

Although telecommunications companies have long followed the rationale that keeping networks out of public view would increase their security, today this invisibility poses a threat to the cables themselves and at times to the people who use them.[19] If cables remain invisible to policy makers, government regulators, corporate customers, business managers, and politicians, then critical decisions about infrastructure funding—which could make our networks more robust and accessible—will continue to be uninformed. John Hibbard, president of the Pacific Telecommunications Council, recounted the comments of a local regulator at a cable meeting in Singapore: "Why am I here?" the man asked. "Everything comes into the country via satellite."[20] The regulator's lack of knowledge was stunning because, as Hibbard quipped, "Singapore is about the most wired country in the world. The only reason it doesn't move is because it is tied down by all of these undersea cables." The lack of awareness extends even to the highest levels of the U.S. government: President Barack Obama's 2013 executive order on cybersecurity made no specific mention of the undersea cable industry.[21]

The invisibility of cables also frustrates the industry in its attempts to gain protection or development rights from nations and state-run agencies. Fiona Beck, CEO of the Southern Cross Cable Network, told me that much of her time with investment bankers and regulatory bodies is spent dealing with the question "Isn't satellite bigger and faster and newer than cables?" and that this is an enormous block to getting better legislation.[22] As the uses of coastal and marine space have intensified, cable companies have had conflicts with fishermen and boaters, environmental advocates, and local developers, all of whom need to be informed of cable routes in order to avoid them.[23] Perhaps most significantly, millions of Internet users around the world rely on undersea cable systems for social, political, economic, and media exchanges, but have little recognition of the structures of dependency into which they are often locked. When cables are built, sold, disrupted, upgraded, and rerouted, these changes have significant consequences for their own use of the Internet.

As it traverses the material environments of cable systems, *The Undersea Network* introduces readers to the structure of cable networks, the geographies from which they have emerged and remain sedimented, and the actors responsible for their construction. In the process, the book develops a view of global cable infrastructure that is counterintuitive yet complementary to the popular understanding of networking. It is wired rather than wireless; semicentralized rather than distributed; territorially entrenched rather than deterritorialized; precarious rather than resilient; and rural and aquatic rather than urban. It is my hope that this alternative representation will give digital media users not only an understanding of their own position in a spatial and environmental Internet, as well as of its extraordinary costs, but also a ground from which to argue for new kinds of structures.

From Distributed to Semicentralized

Contemporary networks are often imagined as a distributed mesh, in which individual nodes are multiply linked in an amorphous and flexible topology.[24] These distributed systems are not simply opposed to centralized structures, but, as Alexander Galloway has noted, the "distributed network is the new citadel, the new army, the new power."[25] Indeed, the Internet's decentralized routing system often appears to be the prime example of this technological transition. From the perspective of global cable infrastructure, however, the actual geographic dispersion of signal paths is relatively limited, and the paths remain centralized in key locations. Only forty-five undersea cables extend outward from the continental United States, supporting almost all of the country's international data transactions.[26] If one groups the cables into thirty-mile stretches, one can see that international traffic enters the United States through fewer than twenty zones. This number is high if one looks around the globe. Many countries have less than five external links. This concentration of cables is partially due to the enormous capacity and expense of each system. One recent cable that connects Australia and Guam has enough capacity to carry simultaneous phone calls from the entire population of Australia—over twenty million people.[27] Large transoceanic projects might cost upward of a billion dollars. Since capacity requirements have already been met in much of the developed world, there is not always an economic incentive to diversify one's infrastructure. The concentration is also due to the traction of existing networks: there are a limited number of sites that are insulated from harmful shipping traffic and where one can interconnect with existing systems. Almost all of Australia's Internet traffic goes out through a single thirty-mile stretch, in part thanks to the cable protection zones that insulate that path.

The geography of the undersea cable system is not a distributed network in which all points easily connect to all other points. Rather, it looks more like Paul Baran's description of a decentralized system, in which there are several nodes that are all connected to a central hub and, at times, to one another.[28] When one overlays considerations of control—since a single company might be in charge of all gateways from a country—the network's geography moves further from a decentralized or distributed ideal. It is difficult to glean this from a traditional network map, which shows multiple logical pathways between endpoints but fails to reveal where these pathways use the same physical route, form bottlenecks at narrow points, or are owned and operated by a single company. In directing attention to the relative centralization in the geography and operations of cable systems, this book contributes to emerging research that documents how centralizing forces continue to permeate and underpin the extension of networks—from the U.S. investments in the Cold War to Google's domination of online searching.[29]

From Deterritorialized to Territorial

Following from the popular imagination of wirelessness and dematerialization is a common assumption that digital communications are being freed from territorial limitations. *The Undersea Network* demonstrates how cable routings have been critically shaped by territorial politics and how established political ties have facilitated the development of international communication. Early telegraph networks were mapped over colonial geographies, and the majority of companies that laid telephone cables through the 1980s were government-owned or -affiliated monopolies. These extensive investments shaped the contours of cabled environments and provided traction for Internet infrastructure. The two fiber-optic cables connecting New Zealand to the outside world, for example, are located in the same zones as telegraph cables from the early twentieth century. Takapuna (on Auckland's now suburban north shore) and Muriwai (on its rural west coast) have been landing points since 1912. Major transpacific cable hubs in the United States are located at sites established during the Cold War.

Although our digital environment appears to be a space of mobility, radically changing every few years, the backbone for the global Internet continues to be sunk along historical and political lines, tending to reinforce existing global inequalities. This geographic stasis is also a reflection of the conservative nature of the cable industry: cable technologies are designed to last twenty-five years, installation techniques have changed little since they were developed, and engineers tend to err in the direction of what has already been tested and proven.[30] Advocates for new systems are attempting to remedy this imbalanced geography. Recent moves to network previously uncabled locations, including the Interchange cable to Tonga; the Honotua cable to Tahiti; and the BRICS cable linking Brazil, Russia, India, China, and South Africa, often depend on existing territorial alliances and national governments for funding. Rather than making geography matter less, cable networks continue to be constructed in a dense web of existing territorial affiliations.

From Resilient to Precarious

Packet-switching technology, which forms the basis for the Internet's distributive operations, is often understood in terms of its potential to survive an attack: if several nodes in a network are disrupted, the system's routing can in theory move traffic around them. The relative centralization of the cable system and its embeddedness in existing territorialities make the physical networks across which these packets move far less resilient than we imagine.[31] In 2006 an earthquake near Taiwan set off an undersea landslide and snapped several cables. Another significant outage happened in Vietnam in 2007, when cable thieves pulled up several of the country's working lines. In 2011 a woman in Georgia shut down much of the Internet in Armenia when she dug up two fiber-optic lines while looking for scrap metal. Although network carriers are often able to reroute traffic, in many cases, breaks have decreased Internet connectivity. Repairing undersea networks is dependent on a limited number of specialized cable ships, and in some places, including China, Italy, and Indonesia, companies have had to wait to receive permits before they can fix their systems.[32]

Looking at moments of actual and imagined interference, this book increases awareness about the vulnerability of our networked systems. It would not be difficult for the U.S. government (or any government) to physically switch off all international telecommunications. This is not an imminent possibility, as it would cause extraordinary economic harm that would outweigh any political benefits. As one study succinctly observed, "The entire global economy relies on the uninterrupted usage of the vast undersea cable communications infrastructure."[33] To separate oneself from this economy would be disastrous for most countries, yet this is nonetheless a possibility built into our current system. This book views the narrow points where cables run together as pressure points, sites where currents can be diverted or rerouted using minimal force and where local actors have a disproportionate amount of power. This might occur not only via technological disruption, but also in the mismanagement of the banal dimensions of maintenance and upkeep, long a blind spot in studies of global and digital media. As Stephen Graham and Nigel Thrift argue, social theory has broadly focused on connection and assembly to the exclusion of the "massive and continuous work" needed to keep infrastructure systems in operation.[34] Cable systems are thus also vulnerable from within: an immense amount of time, energy, and embodied labor are required to sustain undersea networks, and without this labor, the infrastructure would soon fail.

Although the Internet is often imagined as a clean and durable technology, something that will eventually be extended everywhere at little cost, this vision fails to register the extensive financial, social, and environmental investments required to establish new systems and maintain existing ones. Taking this into consideration, we might think about the Internet not as a renewable resource but as a precarious platform, especially as moving our data to the cloud often entails increased dependence on undersea links. On one hand, this might help us make better choices about our own media consumption and content production, taking into account the potential precariousness of infrastructural systems. On the other hand, it might motivate us to push for a more genuinely distributed, resilient, and equitable network.

From Urban to Rural and Aquatic

Geographies of digital media tend to focus on the city as it has been intertwined with the development of information flows.[35] Indeed, the destination of signal traffic is often the urban user, and the city has exerted a gravitational effect on infrastructural development. Most of the undersea cable network's routes and pressure points, however, are nestled in natural environments, and the system has been profoundly shaped by the politics of rural, remote, and island locations. Much of Australia and New Zealand's cable infrastructure is routed through and shaped by the histories of Hawai'i and Fiji. A significant amount of U.S.-Asia traffic moves through Guam. On California's west coast, traffic often exits the country via remotely located hubs in San Luis Obispo and Manchester rather than Los Angeles or San Francisco. As a result, the local investments of environmentalists in California, fishermen in Southeast Asia, and deep-sea marine biologists in Canada have come to inflect cable networks in unexpected ways. Although cable traffic is often destined for larger urban areas (very little material drops off in these remote locales), the channels through which it flows nonetheless depend on investments in and reorganizations of aquatic and coastal environments—sites that have rarely been studied in relation to media distribution. As the ocean becomes subject to increasing spatial pressures, with the acceleration of shipping, underwater mining, and alternative energy projects, such environmental negotiations will continue to be integral to network development.

Turbulent Ecologies

The Undersea Network connects the evolution of cable systems with its shifting material contexts, including not only cultural practices and political formations but also atmospheric, thermodynamic, geological, and biological processes, to expose the complexity that goes into the distribution of digital media. Although early discussions about digital systems focused on their distance from the real world, over the past ten years the rise of spatially embedded systems, digital navigation, and ever-accelerating media obsolescence has drawn attention to the imbrication of digital media in its surrounding environments. Researchers have opened the black boxes of digital storage technologies, sifted through the depths of code, located hidden data centers, unearthed the electrical systems that sustain media production, and examined the materialist energy of media systems.[36] Media archaeologists have dug into history to reveal the early predecessors of digital media in predigital technologies. Scholars have documented the importance of media infrastructures and distribution—from satellites to ubiquitous computing—and revealed the lasting effects of technological networks on today's information circulation.[37] Others have examined the specificity of fiber-optic systems, unpacking their logics of control and freedom.[38] *The Undersea Network* builds on this materialist research to document "the physicality of the virtual."[39]

Tracking these changes throughout history, the book develops a new approach—what I describe as network archaeology—to historicize the movements and connections enabled by distribution systems and to reveal the environments that shape contemporary media circulation. Based on existing research in media archaeology, a network archaeological approach draws on archives and historical narratives to shed light on emerging practices and, in light of these practices, to offer new vantage points on the past.[40] To do so, *The Undersea Network* follows the paths of our signal transmissions—from the cable stations in which signals terminate, through the zones in which they come ashore, and to the deep ocean in which they are submerged. These zones, obscured in the thin lines of the network diagram, are the material geographies of cable communications, and through their excavation we can begin to understand the semicentralized, territorial, precarious, and rural natures of digital networks.

To better illustrate what this entails, let us turn to the discussion of a recently proposed undersea network, the Arctic Fibre system (figure I.4). Over the past decade, as the Arctic ice has retreated with global warming, the Northwest Passage has opened up new pathways, not only for shipping and for oil extraction, but also for cables carrying digital communications signals. The proposed Arctic Fibre cable would link London and Tokyo via the Arctic Ocean, a shorter path than the Atlantic and Pacific routes, and provide a new source of Internet connectivity for northern communities.[41] There had been a number of attempts to lay a transarctic cable prior to this, including a telegraph stretching between Alaska and Russia (before the transatlantic telegraph was laid in the 1860s) and Project Snowboard, initiated by British Telecom in the 1980s. In the 1990s the Russian Ministry of Posts and Telecommunications

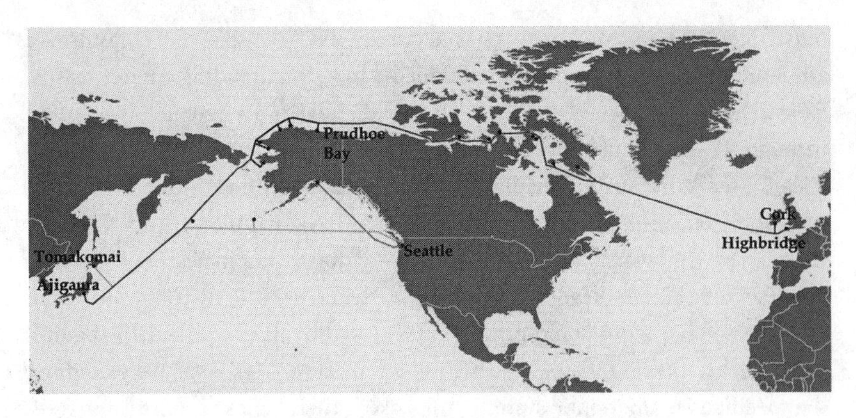

FIGURE I.4. Route of Arctic Fibre's northern cable, January 2014. Courtesy of Arctic Fibre.

even devised a plan to use a nuclear submarine to lay a fiber-optic cable under the Arctic.[42] It was not until the large-scale environmental transformation of climate change, however, that a transarctic cable route became feasible.

Between the route's endpoints in London and Tokyo lie a disparate set of environments—frigid Arctic waters in which deep oil reserves are nested, Canadian and Alaskan communities, and locations where scientific research on global warming is being conducted. Through these predominantly rural environments extend a range of human and nonhuman circulations, from atmospheric currents to the movements of container ships. Such circulations could generate friction for Arctic Fibre, a form of resistance that Anna Tsing describes as simultaneously productive and enabling.[43] The reactions of previously uncabled populations, from the indigenous people of the Canadian north to the oil companies that seek to drill off the coast, are still unknown. Icebergs scouring the coastal seafloor might disrupt shallow cables. Fishermen's nets threaten to hook and sever them. Even along well-traveled routes, environments have always generated friction for undersea networks. Throughout the telegraph era, fishermen regularly dragged their nets along the densely cabled transatlantic route, disconnecting links and scattering signal traffic.

These circulations generate interference for the system. For Arctic Fibre, the movement of ice not only threatens to break cables; in the deeper sea, it also covers much of the ocean's surface, literally interfering with the company's access to their network. In other areas around the world, cyclones, tectonic plate shifts, and rising waters threaten to physically disrupt the movement of media and communications. Interference can also be created by social and cultural phenomena and, at times, is generated as a by-product of the cable's extension. When some telegraph companies constructed cable stations in remote colonies, tensions were generated with indigenous people who would later resist communications development. The environments that cables stitch together are not always smooth spaces, but turbulent ecologies. Turbulence is a chaotic form of motion that is produced when the speed of a fluid exceeds a threshold relative to the environment it is moving through. Not an uncommon occurrence, turbulence is the "rule, not the exception, in fluid dynamics."[44] When a fluid—whether air, water, or blood—becomes turbulent, it breaks down into smaller swirling currents, called eddies, which in a cascade break down into smaller and smaller irregular flows. Turbulence is rarely a direct and purposeful opposition to flow. Rather, it describes the way that social or natural forces inadvertently create interference in transmission simply because they occupy the same environment, in the end contributing to the network's precariousness.

Cable companies go to great lengths to protect against both real and imagined forms of interference. In order to facilitate smooth and reliable signal exchanges, they develop extensive social, architectural, geographic, and discursive strategies of insulation. In this book, I define *strategies of insulation* as modes of spatial organization that are established to transform potentially turbulent ecologies into friction-free surfaces and turn precarious links into resilient ones. All along transoceanic cable routes—at the cable station, the cable landing, and in the deep sea—cable owners, manufacturers, and investors reorganize these spaces in order to enable the continuous flow of electrical and political power. A strategic organization of space, as Michel de Certeau notes, "becomes possible as soon as a subject with will and power (a business, an army, a city, a scientific institution) can be isolated . . . every 'strategic' realization seeks first of all to distinguish its 'own' place, that is, the place of its own power and will, from an 'environment.' . . . It allows one to capitalize acquired advantages, to prepare future expansions, and thus to give oneself a certain independence with respect to the variability of circumstance. It is the mastery of time through the foundation of an autonomous place."[45] Strategies of insulation are designed and financed by companies to distinguish the spaces of distribution networks as "an autonomous place" and to separate them from conflicting circulations. Approaching a historically uncabled environment, Arctic Fibre will have to fund such strategies of insulation, such as the burying or double-armoring of the cables at problematic sites, the establishment of new "no anchor" zones to keep fishermen out of the cable area, and the monitoring of icebergs via satellite.[46] Whether they are social, architectural, or discursive arrangements, strategies of insulation separate one part of an environment from the rest to stabilize the distribution of media and communication.

Local and regional circulations are not always disruptive to cable systems; in many cases, networks are planned so as to incorporate them into the circuit. For Arctic Fibre, potential users of the cable system are seen as untapped economic circulations: the regional practices and investments of the Canadian High Arctic Research Center Service, Canada's Department of Defense, and oil companies that seek to drill in the area will help to fund the network.[47] Douglas Cunningham, Arctic Fibre's CEO, argues that revenue from international traffic passing between London and Tokyo cannot in itself fund the project, and that any northern project requires either domestic demand or a government subsidy—in effect, funding that is generated in the environments on the cable's route, rather than simply at its endpoints.[48] Although coastal communities' Internet use will never financially sustain the cable alone, serving these communities also remains integral to Arctic Fibre's pitch for funding from the Canadian government, since the cable could lower Canada's costs to provide health care, education, and other government functions to the region. These interconnections highlight the continued roles of territoriality and nationalism in supporting cable networks.

In appealing to these users, Arctic Fibre seeks to develop new strategies of interconnection, modes of spatial organization that are designed to leverage local and regional circulations, or at least perceived circulations, to support its cable network. This might entail setting up actual technological points of interconnection, or gateways, where signals can be transferred between networks. Technical interconnection is an important concept in cable management: cable networks are solidified in particular locations via formal interconnection agreements, where competing companies build bridges to each other for their mutual benefit. Although the process of interconnection, critical for all kinds of networks, has been described in various ways, I use the term *strategies of interconnection* to refer to the development of fixed architectures and spatial practices through which transfers between the cable system and its surrounding environments can occur.[49] Arctic Fibre will develop strategies of interconnection to link not only to potential users of the network, but also to the existing resources in the Arctic and oceanic environment, including icebreaker ships and remote-operated vehicles that could aid in repairs. Such strategies are designed to facilitate the process of transduction—the transfer of energy, whether social, economic, biological, physical, or electric, between the system of the cable network and the cultural geographies into which it is inserted.[50]

In these cases, the network's proximate environments are not a site of interference, but a critical link in the construction of global communications, a place where signals are grounded. In electrical engineering, the movement of a current from one point to another creates an imbalance in transmission, and these circuits are made stable only by a return channel, which is termed the *ground*. Technically, for undersea cables, the ground is the ocean itself. A signal moves through the cable across the sea, and then the energy returns to its original location via the conduction of the water. As one cable report observes, powering the operation of a system "is achieved by actually using the environment—the seawater—as a conductor to complete the circuit."[51] Grounding is thus a process whereby the conductivity of existing matter—whether made up of social or natural phenomena—is harnessed to keep a current from becoming imbalanced. This metaphor highlights the fact that strategies of interconnection are not simply modes of exploiting the environments in between nodes, but also serve an important role in keeping a network in equilibrium. Today, interconnection is often facilitated by technologies, but throughout history this process has been a thoroughly human endeavor. For example, in the early remote cable stations, imbalances were created as young men moved to locations far from home and were subsequently forced to migrate between stations. Slowly the company and the cablemen developed strategies of interconnection, ranging from using local labor to arranging marriages, which helped to sustain the operators and therefore stabilize transoceanic signal traffic in remote locales. Through interconnection, energy from one system was transduced to support another's deficiencies, and equilibrium was achieved (or at least attempted).

The concepts introduced here—strategies of insulation, designed to transform turbulent ecologies into friction-free surfaces, and strategies of interconnection, designed to ground transoceanic currents in local circulations—describe the dynamics by which the infrastructures of digital media are formed in relation to their environments. Strategies of insulation shelter the movement of international signal traffic from the environments they traverse. They produce an internal break in an ecology, allowing one system to extend into and through another without being affected by it. Cables must be insulated from hurricanes and fishermen, local publics and foreign nations. This practice of intentional disconnection is integral to sustaining and securing network operations in potentially turbulent environments, and it has intensified rather than subsided as networks have become more critical to our global society. However, at the same time, the network must be grounded in some way via strategies of interconnection, the leveraging of existing circulations to support new networks. Although they appear to have opposing relationships to the environment (blocking it out versus harnessing it), insulation and interconnection are complementary strategies that regulate cable ecologies and stabilize circuits of transmission on a global scale.

Insulation and interconnection—from the establishment of cable protection zones to the channeling of local labor—reduce the threat of disruption to the cable system and make cable laying less expensive and more efficient along existing routes, a process often described as "path dependence" in studies of technology and social practice.[52] I use the term *traction* to refer to the ways that these interactions—in which cables both repel and connect to preexisting currents—anchor infrastructures in particular sites. When there have been opportunities to build new stations and remake the geography of the cable system, existing routes were typically chosen, because paths through those sites had already been negotiated. The fluidity of our information sphere is made possible only by this historical fixity of communications infrastructure. As David Morley observes, "Right at the heart of the process of globalization, somewhat counter-intuitively, we find some rather important things slowing down."[53] These slow, fixed, and disconnected spatial practices are the hidden layers that support contemporary global networks.

The challenges that Arctic Fibre faces—even at a time when distance seems to matter less and less—entail strategically developing a new path through the Arctic environment. The company must fund the acquisition of new knowledge about an uncharted aquatic and coastal landscape as well as the development of new modes of cable protection. It must generate new connections with existing actors along the cable route and cultivate new markets for cable service. The absence of networks in the area provides the cable layers with a number of challenges that their transpacific and transatlantic competitors avoid. Their ongoing costs will be high, especially since they cannot share maintenance agreements with other systems. The industry will perceive Arctic Fibre as more risky, given that the route has been as yet untested. The lack of historical precedent, all in all, will make it difficult for the company to find funding. If it succeeds in establishing the first transarctic cable, setting up critical installations, and pioneering new modes of protection, it will then be much easier and less expensive for other companies to set up shop along the same route. It is no surprise, then, that most companies build systems along routes where cables have been laid, stations constructed, "no fishing" zones established, and markets formed.

The case of Arctic Fibre, as it plans an innovative route across uncabled waters, forcefully demonstrates how networks take shape in and in turn inflect the environments around them: these are much less pressing concerns in the development of other global routes where surrounding ecologies have been managed for over a century. These ecologies, which consist of social practices, built architectures, and natural environments, have been invisibly folded into the thin lines of each network edge and the production of our intercontinental cable system. They remain hidden, however, in the common cable map. *The Undersea Network*, as it analyzes the historical negotiations and emergence of such ecologies, reveals that the creation of a stable circuit of transmission (more than simply the exchange of a single message) is always an environmental process. It involves manipulating space, from the sediment of the seafloor to the housing options of colonial cablemen, to mold contours across which signals can repeatedly and reliably move without disruption. The concepts developed here—turbulent ecologies, pressure points, strategies of insulation and interconnection, and traction—attune us to these processes, placing the geography of network infrastructure in relief.

The Undersea Network offers a new way to look at digital media systems in ecological terms. Although research on media ecology—from Neil Postman's studies of mass media to Matthew Fuller's *Media Ecologies*—understand the environment as a world of content, *The Undersea Network* extends the environment to encompass the social, architectural, and natural ecologies through which this content is distributed.[54] Here, global information flows are not positioned as equalizing, deterritorializing, and antithetical to fixed or hierarchical structures, but instead are always routed through dynamic fields made up of varied directional circulations. The challenge for networked circulations today is not how to overcome fixed barriers, but how to navigate in a world where everything is already mobile. The title of this book, *The Undersea Network*, thus refers not simply to the cables that are being analyzed, but to the book's methodological intervention: to see networks as always embedded within complex and multidirectional circulatory practices—not a static territory, but a fluid environment in which our connections must be both insulated and grounded.

Vectors

To excavate the formative role of these environments, I followed the undersea cable route.[55] This brought me to cable stations, where I interviewed telecommunications workers about cable operation and maintenance, and to landing sites, where I spoke with residents about their encounters with cable networks. I visited industry conferences and interviewed people in engineering and in marine operations as well as in sales, marketing, finance, and law about their experiences in setting up cable systems. The industry has a reputation for both secrecy and speculation; unraveling its history has involved sorting out myths and rumors and has taken me to local, national, and corporate archives.[56] As a corrective to what I see as the fundamentally limited visibility of cable systems, I have photographed cable networks throughout my travels to develop new approaches to cable representation. Cumulatively, this combination of ethnographic, archival, and artistic fieldwork offers a multivalent model for studying distribution systems, a network archaeology that connects cables' historical and technical organization to the layered cultural, political, and biological environments that surround them.

In this book, I hope to convey how the network looks to the people who build, operate, and use it. I visited cable installations in thirteen countries, as well as the offices and homes of numerous cable workers who have contributed to these systems. As a white American woman, I was unfamiliar with the culture and language in many of these places, and I relied on people who were in the industry and resided in these geographies to make the systems legible. Almost always, it was men—an analyst at a cable company in New Zealand, a distant relative in Guam, and a software developer in Yap—who broadened my mobility and helped me gain access to cable networks. While the cable industry is overwhelmingly male, many women have also been involved in the industry, particularly in labs and in project management, sales, marketing, route surveying, and legal affairs. To get to cable sites, I also relied extensively on the infrastructures of transportation (roads, boats, planes, and trains) with which the cable systems have been interconnected. My own subject position—as I traversed heterogeneous infrastructures and environments, often feeling out of place—heightened my perception of cables' ongoing requirements for interconnection as well as their vulnerabilities and need for insulation.

Each of the following chapters focuses on a distinct environment in which cables have taken shape. The first chapter, "Circuitous Routes: From Topology to Topography," gives a broad overview of the three major eras of cable development, providing the backdrop for the rest of the book. In it, I sketch out the large-scale cultural forces that affected cable networks, beginning with the copper telegraph cable's relationship to colonization from the 1850s to the 1950s, extending through the coaxial telephone cable's imbrication in postwar politics from the 1950s to the 1980s, and ending with emergence of fiber-optic cables in relation to deregulation and privatization from the 1990s on. To understand the routing of these networks, I argue that we need to move away from network topology, the analysis of the mathematical structure of connections, to topography, the analysis of how cables have been embedded into historical and geographic matrices. Counter to typical assumptions, cable geographies do not simply follow a terrestrial logic (they are often laid underwater when possible), an urban logic (they often connect suburban or rural environments), or a demand-driven logic (they often connect places that are already connected). Rather, I show how over the past hundred years transpacific cable systems have been constructed to be secure: they are deliberately routed to insulate signal flow from potential sources of natural and social interference, from a nuclear bomb to a terrorist attack. Cable routing is also driven by a competing tendency to interconnect: systems have often been routed in inconvenient and expensive ways in order to link with other systems. On the whole, this dynamic gives new routes traction in existing topographies, leaving us with a relatively centralized global network.

The second chapter, "Short-Circuiting Discursive Infrastructure: From Connection to Transmission," follows cables into the discursive environments of popular media. I argue that almost all stories about undersea cables fit into one of two narrative modes. Connection narratives trace the development and initiation of the cable, aligning this event with a transcendence of national boundaries and the easing of international conflicts. Disruption narratives focus on a cable's repair after it has been disconnected, narrating the event as a fight against broader threats to global connectivity—including nature, nations, and terrorism. Both of these narrative modes are limited: they depict the cable only when it is out of service and, as a result, exclude the enormous amount of work involved in the upkeep of global systems. Rhetorically, they function as strategies of insulation that have, until now, protected cable systems. The chapter delineates two alternative approaches that represent undersea cables as material infrastructures: nodal narratives, which focus on a node in the system and chronicle the human and nonhuman extensions through it, and transmission narratives, which move with a signal as it is transmitted through the cable. By narrating the cable past the moment of its initiation, they extend the spatial and temporal parameters for cable discourse, suggest new lines of causality in global network development, and set the groundwork for further engagement with operational cable systems. I argue that, in doing so, these narratives short-circuit the ideological power conducted by narratives of connection and disruption.

The third chapter, "Gateway: From Cable Colony to Network Operations Center," details the history of the cable station as a gateway to the network: it is a site of interconnection between national and international systems, a place where connections are made to local publics, and a zone where the border between system and environment is contested. The chapter moves through each of the three periods of cable development, documenting the shifting boundary between stations and their surrounding ecologies. In the colonial cable station, the cable worker's body was the crux of network operations and the zone to be protected and regulated. As stations were remade during the Cold War, the border between the network and environment shifted from the body to the station's built architecture. In the fiber-optic era, strategies of insulation now regulate the circulation of information. In each period, I highlight the investments in insulating the station and demarcating the inside and outside of the network, alongside the strategies of interconnection that ground the system in local micro-circulations. Looking at the network's shifting interface with local publics, the chapter also illustrates how labor and a cable community remain key support systems for information networks.

The fourth chapter, "Pressure Point: Turbulent Ecologies of the Cable Landing," analyzes conflicts at the cable landing, the zone where undersea cables emerge from the deep ocean and extend through coastal waters, beaches, and local communities before connecting with cable stations. These public spaces cannot be walled off and often become pressure points, sites where local actors can induce turbulence in the system. This chapter documents the strategies of insulation developed by cable owners, manufacturers, users, and investors that affect the cable's visibility to the publics who inhabit the landing point. Tracking these interactions in Hawai'i, California, and New Zealand, I describe how small-scale circulations at the cable network's pressure points have produced disproportionate effects across the network.

The fifth chapter, "A Network of Islands: Interconnecting the Pacific," charts how network nodes are shaped by the politics, histories, and geographies of islands across the Pacific. Although existing representations of islands and networks reinforce a conceptual opposition between the two, making it difficult to see both the interconnectedness of islands and the importance of network maintenance, this chapter recasts islands as core components of cable systems. It focuses on three critical points, past and present, in transpacific traffic: Guam, critically tied to American military extensions; Fiji, a key site for British colonization of the Pacific; and Yap, a former node in the German cable network. In these cases, I show how networks have benefited from the island's insulating properties and, in turn, how islands have become sites of interconnection, places where reciprocity can be established between oceanic, cultural, and communications currents. I argue that emerging as a network hub, becoming more than an endpoint for signal traffic, has involved triangulating existing sets of circulations, whether transpacific, regional, or local.

The last chapter, "Cabled Depths: The Aquatic Afterlives of Signal Traffic," analyzes how undersea cables exert a lasting influence on our knowledge about and inhabitation of the ocean. The chapter documents the relationship between early marine science and telegraph cable networks, which together helped chart a distinct set of transoceanic paths. During the era following World War II, these exchanges were increasingly shaped by the U.S. militarization of the seafloor, and cables took on a new function as they were mobilized for the acoustic monitoring of marine space. Today these systems feed back into the development of marine scientific research via the construction of cable-linked ocean observatories, and into new extractive relationships with the seafloor. Although the first five chapters focus primarily on the manipulation of physical sites and social practices, chapter 6 explores the institutional and epistemological interconnections that have inflected cable development. In describing how our knowledge of the ocean is thoroughly intertwined with cable histories, the chapter—like the ones before it—reveals the porous boundary between communications technologies and their environments.

These chapters offer a set of nodal narratives that illustrate the long-standing relationship between media infrastructures, environmental processes, and cultural history. Together, they show how cable companies have developed extensive strategies of insulation for network infrastructure and solidified pathways through social and natural ecologies. This process, involving both the production of knowledge about and the physical reorganization of cables' environment, has often been made possible by large-scale investments—colonial, military and corporate. At the same time, the development of transoceanic signal exchange has also involved leveraging and connecting with local and regional circulations. Weaving through these diverse geographies, the book introduces a sense of place and an environmental consciousness to our imagination of digital networks, prompting consideration of their costs, whether financial, architectural, or social. Rather than being driven by the physics of entropy—where movements are becoming ever more chaotic and interdependent—*The Undersea Network* reveals how that experience of wirelessness is accompanied by an increasing investment in wires; intercontinental connections paradoxically require numerous forms of disconnection; and our experience of global fluidity is made possible by relatively stable distribution routes that perpetuate conditions of uneven access along lines established a century ago. This book charts the movements and channels that push back against flow and ultimately shape the conditions of possibility for circulations across and under oceans.

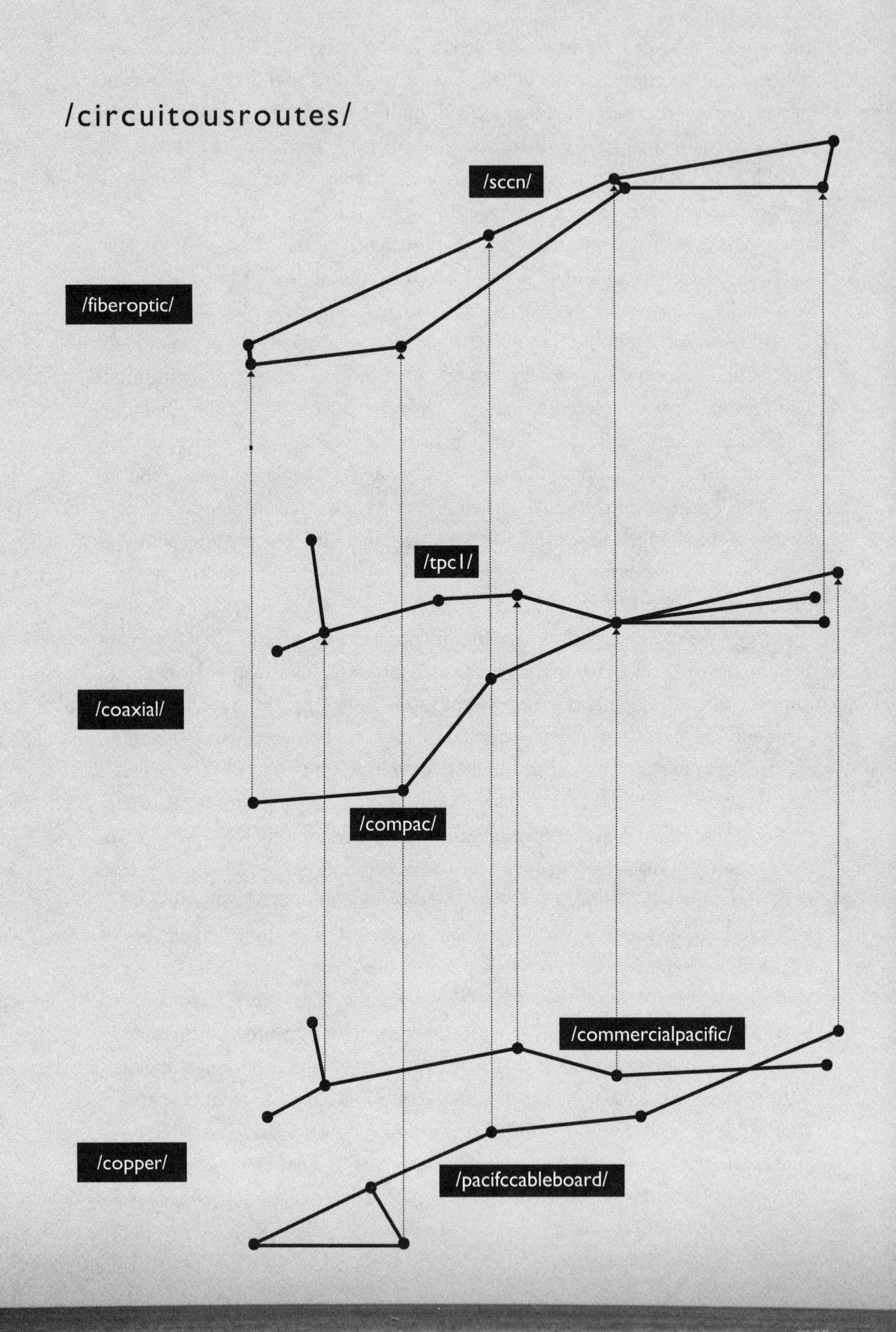
/circuitousroutes/
/sccn/
/fiberoptic/
/tpc1/
/coaxial/
/compac/
/commercialpacific/
/copper/
/pacifccableboard/

ONE

circuitous routes

From Topology to Topography

At first glance, the Arctic Fibre cable system—the 15,600-kilometer network that would link England and Japan via the Arctic Ocean—makes little sense. With plenty of terrestrial networks spanning North America, why pioneer a circuitous route through one of the least populated areas of the world and an expanse of inhospitable environments, where the cable would be covered by several feet of ice during much of the year? Why spend a projected $640 million to connect Arctic communities with a high-capacity cable when they will likely never use a significant percentage of its capabilities? Given the extensive number of data connections and surplus capacity between London and Tokyo, the cable's two endpoints, why does the world need another cable? Similar questions could be posed to other cable projects seeking to spend millions to link already connected endpoints, from the Hawaiki cable, a proposed system between New Zealand and the United States, to the South Pacific Island Network (SPIN), which was designed to hop from island to island across the Pacific.

One answer to these questions is that these cables offer diversity, a redundant and geographically disparate pathway that would make signal traffic more reliable. If established, these systems (or so their projectors argue) would better insulate our information flows from people, environments, and other forms of interference encountered by networks. Arctic Fibre avoids pressure points on existing routes between Europe and Asia, including the earthquakes of the Luzon Strait, South China Sea, and western Pacific, as well as social and political unrest that make the Suez Canal and Mediterranean particularly turbulent environments. It satisfies a need for routes between Europe, the Middle East, and Asia that skip the United States, an option desired by Asian and Middle Eastern carriers who want to secure their traffic from potential monitoring on U.S. territory. It does not hurt, of course, that Arctic Fibre would also be the quickest route from London to Japan, at 168 milliseconds for a transmission, enabling the company to leverage the market of high-frequency traders who seek to capitalize on the path with the shortest transit time.

The diversity of our global networks, long a concern of telecommunications companies, has recently come to the forefront of policy makers' attention. Prompted in part by a series of very public cable failures, representatives of the telecommunications industry, members of the financial sector, technical experts, and policy agents came together in 2009 at the Reliability of Global Undersea Communications Cable Infrastructure Summit (ROGUCCI) to discuss how to make global network infrastructure more resilient. The summit's participants concluded, among other points, that even though individual systems were highly reliable, on a "global level, the overall interconnectivity of the continents violates a fundamental reliability design principle—avoid single points of failure."[1] Participants pointed out several "geopolitical chokepoints" where cable paths were funneled together and warned that a disaster in one of these areas "could cause catastrophic loss of regional and global connectivity."[2] Our global network, with no overarching body to ensure its robustness, has been constructed in an ad hoc way by companies with a range of different interests, creating an infrastructural reality that runs counter to the common imagination of the global Internet as a distributed mesh, able to easily circumvent any attack in a specific geographic region and route traffic around any disruption.

This chapter argues that in order to understand the geography of signal traffic, we must move beyond network topology, the observation of the geometric or mathematical distribution of nodes and links, to consider network topography, the way that infrastructures are embedded into existing natural and cultural environments. Paying attention to network topography counters several widespread assumptions about the determinants of cable routing. First, there is a terrestrial assumption: people believe that undersea cables are used only when a terrestrial cable route is not available. Researcher Linda Main argues that since running cables over land is less expensive, "undersea cables have traditionally been used only between continental landmasses, where terrestrial links are not feasible."[3] Cables are assumed to follow the shortest geographic route between continents in order to minimize underwater segments. When cables do not land at the closest terrestrial points between landmasses, their geography is often explained by an urban logic that assumes that cables directly connect the centers of major cities (reflecting a broader focus on the city as the critical unit of analysis for telecommunications infrastructure).[4] A third assumption is that cable routes are driven by demand: they are extended to places where there is a surplus of signals and not enough cables to carry them. This is an anthropocentric and equalizing view of the cablescape: we assume that, like people, cables must be easier to sustain in accessible and central environments and are simply extended to users who lack connection.

This chapter shows that even though terrestrial, urban, or demand-driven logics may at times play a role in determining cable routes, our global infrastructure has been constructed in relation to historically specific social and environmental imaginations. Security, in the broadest sense, has always played a critical role: companies route cables in ways that insulate them from potential interference in their surrounding environments, ranging from natural disasters to anticipated geopolitical friction. Counter to any presumed terrestrial logic, a secure network has most often been an underwater one—territorial politics have made laying routes on land incredibly difficult. Douglas Cunningham of Arctic Fibre argues that the ocean and its ice will serve as a layer of insulation. He commented in one presentation that "60% of cable breaks are from ship anchors and trawls" and observed that "this route will actually be safer than the other routes because we have, on 34% of the length, ice cover for seven months of the year. And that is protection."[5] The deep sea has been the safest zone for a cable because it is farthest from people and subsea engineering is more reliable than terrestrial engineering (components function for decades underwater). The closer a cable gets to shore, the more heavily armored it must be. Although it is true that many cables do connect urban centers (which remain significant endpoints for data flow), landing points are rarely established in the heart of cities. Rural and suburban environments, like the ocean, protect cables from the potential turbulence of human traffic in densely populated areas. The Arctic Fibre cable, though described as going from London to Tokyo, actually lands far outside these cities.

Complete insulation of a cable network is never entirely possible (or profitable). Such insulation would make it impossible for cables to interconnect with the sources of traffic necessary for operation: other undersea cables; domestic fiber-optic systems; and rail, road, and air transport networks. As Stephen Graham and Simon Marvin argue, for this reason new infrastructures are often layered over existing transport and resource systems and rarely develop in isolation.[6] Cables have often been laid to locations where there is already enough capacity as a way of generating competition, stimulating increased signal traffic, and lowering prices. Although many people assume that cable networks are demand driven, advocates of interconnection often make the economic argument that the networks drive demand. Rather than viewing cable laying as a struggle to overcome the ocean or link urban nodes, we might see it as a process of securing routes from turbulent environments and interconnecting them with existing cultural and technological circuits.

Moving from the techniques of the telegraph era, through those of the coaxial period, and to those of today's fiber-optic networks, this chapter describes how investments in cabled environments have shaped the topographies of three generations of transpacific cables and provides a context for the situated geographic analysis of the following chapters. The global telegraph network, constructed in the second half of the nineteenth century, drew support from colonial networks and pioneered the use of the ocean as a layer of insulation. The reinvention of cable technologies after World War II involved negotiations between existing routes of empire, emerging forces of infrastructural decentralization, and a new club system of cable laying (described below). Today, the lines along which the Internet flows evidence a similar push and pull: deregulation and privatization have helped pioneer a new cable geography, which nonetheless is layered into a geopolitical matrix of preexisting colonial and national routes. This genealogy will be familiar to some readers, as historians have analyzed the complex histories of telecommunications systems in detail elsewhere and have well documented the importance of economic and political influences on cable development.[7]

The aim of this chapter—with its focus on security, insulation, and interconnection, especially in the construction of new routes—is to recast our understanding of cable networks to better account for their contingent and material qualities, projecting them as "fragile achievements" that reflect both the ideals of the past and the geographic contours of the earth.[8] I suggest that we view cables not simply as technical systems, but as long-lasting contours in the environment—places where capital, labor, and knowledge have been sunk into the earth's surface, "accumulated imprints" of investment in particular spaces.[9] As a riverbed shapes the flow of water through it, such contours affect—though they do not determine—the direction and force of subsequent circulations. In pursuing the most secure routes, individual cable companies have solidified the network along relatively few lines. As a result, in many places our cable network remains circuitous rather than direct, underwater instead of on land, rural rather than urban, and connects places that are already connected. Perhaps ironically, this tends to perpetuate unequal topographies of global exchange and has left us with a relatively concentrated, semicentralized, and precarious geography that is now proving expensive and difficult—as the case of Arctic Fibre shows—to diversify.

Copper Cable Colonialism

The geography of telegraph routes in the late nineteenth century followed transportation and trade routes, many of which had been pioneered by British colonial investment and served to support existing networks of global business. Cables often landed at the same sites as ships, not only to interconnect with marine transport (the shipping industry was a significant user of cable systems), but also because these sites were often geologically appropriate for both forms of traffic (each required a smooth transition between land and sea). Cables were strewn between ports in the Pacific, from the major hubs of Hong Kong and Singapore to outposts such as Port Darwin. It is no coincidence that Australian cables were brought ashore ten kilometers south of Sydney at Botany Bay, where Captain James Cook claimed the continent for England in 1770 and where the French explorer Jean François de Galaup, comte de Lapérouse, landed in 1788 (figure 1.1).[10] Even at the network's more remote landing points, from Bolinao in the Philippines to Banjoewangi in Indonesia, there was almost always an existing set of colonial infrastructures, however limited, that could support the new stations.

The selection of network routes during this period rarely embodied a completely terrestrial or urban approach. Instead, it represented a balance between the need to interconnect with existing populations and infrastructures and the affordances of an area's natural and social topography. For example, in 1876 the Eastern Telegraph Company laid a trans-Tasman cable from Botany Bay to Nelson, where the company sought to connect with economic trade that was then centralized in New Zealand's South Island. Nelson was neither the closest geographic point to Australia, which was determined to be too remote to support even the cable station, nor a major commercial center—Christchurch was the endpoint for much of this traffic but was deemed too far away. Rather, this route, like many others, was a compromise: Nelson gave the cablemen

FIGURE 1.1. The Botany Bay cable station is adjacent to a monument (far right) honoring the scientist Claude-François-Joseph Receveur, who was part of the first French expedition to the area—a material reminder of the site's many historical landings.

sufficient access to existing infrastructure but did not needlessly waste expensive undersea cable. Landing a cable in Nelson, as was true for other locations such as Botany Bay and San Francisco, entailed laying a longer marine route between landmasses than was absolutely necessary. Although arguments were certainly made about the need to directly connect urban hubs and to minimize undersea routes, it was not uncommon for such statements to be made by individuals with specific commercial interests or geographic pretensions.[11]

More than any single technological justification, security and insulation were important to the negotiation of cable routes. When undersea cables were first laid in the 1850s, the most significant challenge of installation was protecting them from water. Terrestrial wires could remain without physical insulation since air, a nonconductive medium, kept signals from diffusing, but insulation was of paramount importance for an undersea cable. Without it, the signal would easily dissipate into a conductive ocean. In his history of cables, Willoughby Smith dates the origins of undersea cabling not from the successful conduction of signals but from adequate insulation: he considers the 1850 cable from England to France the first pioneer of undersea telegraphy, despite the fact that it conveyed indecipherable messages, because it proved that a circuit of power could be insulated from water.[12] British companies were able to dominate the cable business throughout the late 1800s due not only to their ability to master cable-laying technology and monopolize technical-support infrastructure, but also to their development of an effective form of insulation using gutta-percha, the rubber-like gum from Malaysian trees, and their control over its extraction and distribution.[13]

As the network expanded, aquatic environments once perceived as a threat became the most significant form of cable protection. Counterintuitively, the weakest parts of cable networks were the terrestrial links and coastal segments; in contrast, the deep ocean was the safest zone.[14] Overland cables could become targets during popular revolts, as happened during the Boxer Rebellion in China, when telegraph lines were sabotaged, cutting off communication between local diplomats and the British government. Australia's transcontinental line was often disconnected by Aborigines and the occasional lost cowboy who wandered across the landscape, cutting the line in order to be rescued by cable repairmen. Most natural breaks occurred in shallow coastal waters where tides dragged the cable across rocks. In contrast, very few breaks occurred in the deep ocean, where cables functioned for decades without interruption. To establish political support for the transpacific cables, for example, the argument was made again and again that "in the depths of the Pacific Ocean, the cable would be absolutely safe from interference."[15] In addition to keeping the network safe from perceived physical disruption, running cables through extraterritorial space enabled users to circumvent transit taxes that might have been levied by countries as messages passed through their territory.

Companies began to use the ocean as a layer of insulation to make cables less accessible and visible, thus protecting them from the turbulence of physical, social, and economic conflicts above. Several sites on the Pacific Cable Board's network were initially chosen to minimize undersea segments: the company landed cables at Doubtless Bay instead of Auckland (New Zealand), Southport instead of Sydney (Australia), and Bamfield instead of Port Alberni (Canada) and then linked them via overland lines to populated hubs. However, these terrestrial links soon caused problems. The company worried about the loss of revenue and prestige that could be caused by disruption to New Zealand's "precarious overland route"; the Australian overland line from Southport to Sydney became notoriously unreliable; and the terrestrial line in Canada was often disrupted by winter storms, as well as bush fires and electrical disturbances.[16] The short Canadian segment, not the nearly 4,000-mile-long undersea cable to which it was connected, was referred to as the system's weakest link. Val Hughes recollects: "Getting a message [from Bamfield] to Vancouver often meant sending via Australia and all around the world the other way!"[17] Eventually, more secure undersea cables replaced each of these terrestrial lines.

The global spread of undersea communications networks during this period was led by private companies rather than governments, but it was none-

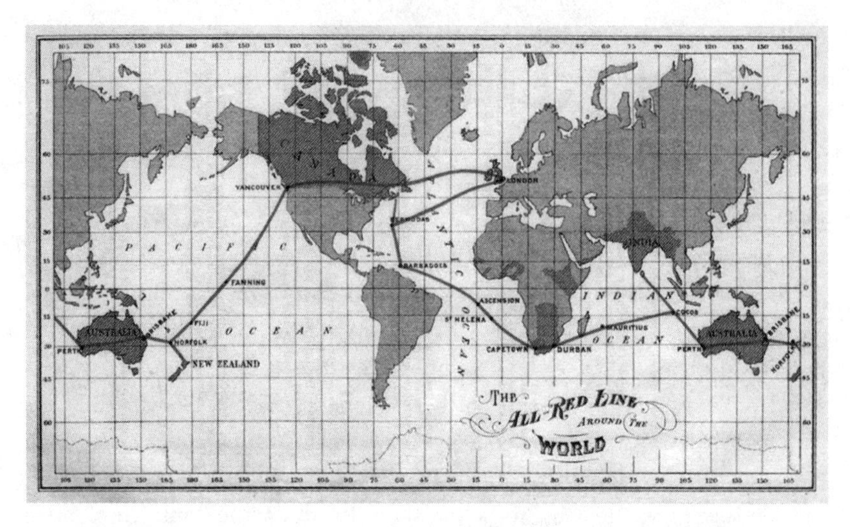

FIGURE 1.2. The All-Red Line, 1902. From George Johnson, *The All Red Line: The Annals and Aims of the Pacific Cable Project* (Ottawa: James Hope and Sons, 1903).

theless framed as a strategy to secure colonial empires. In the latter half of the nineteenth century, the British government supported and subsidized the creation of the All-Red Line, an undersea system composed of strategic cables linking many of the British colonies (some with little commercial value) and landing only on British soil (figure 1.2). Alex Nalbach refers to cables as "the hardware of the new imperialism."[18] Military strategists saw cables as the most efficient and secure mode of communication with the colonies—and, by implication, of control over them—especially during wartime, when enemies might use the geographic dispersion of the empire to their advantage. The invisibility of undersea cables also enabled the extension of these networks (and the perceived control that accompanied them) to places where they might have otherwise been opposed. When there was resistance to the installation of the telegraph in China because it conflicted with customary beliefs (the Chinese government rejected plans to link land telegraph lines to treaty ports), undersea cables were secretly brought ashore in the middle of the night.[19]

In many cases, the effectiveness of undersea cables in actually securing colonies or territorial holdings remained an aspiration rather than a reality. Daniel Headrick writes: "It is tempting to believe that by putting colonies in rapid contact with London, cables gave the Colonial and India offices a tighter reign over their distant subordinates, thus substituting centralized control for the little subimperialisms of the periphery. The evidence, however, points the other way. For information about crises on the frontiers of the empire, London still depended on the men on the spot, who distorted the facts to suit their own ambitions."[20] Leakage occurred in places where diffusion was not expected, at the point of reception where power was diverted to suit the aims of "men on the spot" (a tactic of short-circuiting power rather than opposing it). Along similar lines, in his discussion of the All-Red Line's connection to New Zealand, James Smithies observes that the cable was economically unproductive and prioritized over other pressing basic infrastructure projects, and rather than creating a simple subservience to British rule, it went "hand in hand" with a growth in colonial nationalism.[21] In the end, the extension of the underwater network, even if did not actually increase control over the sites it covered, indexed the belief that undersea cables could secure economic and political investment. This perception played a key role in justifying networks: for example, the Danish Great Northern Telegraph Company was able to acquire funds for a set of cables connecting parts of Europe and Asia in part because of the desire of Russia and other nations to have a secure communications route that ran independently of British links.

Although security in the broadest sense was certainly important in the nineteenth-century telegraph network, the early global cable network supported a globalized media system, rather than simply a colonial one. As Dwayne Winseck and Robert Pike have shown, it was managed by multinational cartels, and its markets operated with transnational financing: this enabled companies to connect to each other with relative ease.[22] At the beginning of the twentieth century, territorial security became an increasingly important rationale for extending networks—it played a key role in the design of the first two transpacific cables.[23] Prior to this time, cable traffic from Europe to Asia and Australia was routed either by way of Singapore (where the Eastern Telegraph Company had a monopoly) or via a transsiberian line (where the Great Northern Company had a monopoly). This geography kept London and Europe at the center of the cable empire, and companies based in Britain resisted any efforts to develop a transpacific link that would diversify the network and potentially reduce their power.[24] Like the Arctic Fibre project, transpacific cables were not understood to be commercially viable on their own, since at the time there was not much transpacific trade, and they would link territories already connected to the network.

As a result, the first two transpacific systems were justified due to the diversity they offered in comparison with existing routes and the perceived security that the added diversity would bring to the system as a whole, especially if the cable route through the Mediterranean was disrupted. A transpacific link was also seen as a way to break up Eastern Extension's monopoly on the links to New Zealand and Australia and to add a speedier route. As Simone Müeller-Pohl has argued, for this reason crossing the Pacific became tenable only once Britain began to be concerned about imperial security.[25] Sanford Fleming and the Canadian government mobilized arguments about security and helped establish the Pacific Cable Board cable in 1902, setting what would be a foundational route for all north-south transpacific traffic. During this period, creating a secure network meant avoiding sites of potential foreign intervention: the final route—which stretched from Bamfield, Canada, to Fanning Island, Fiji, and Norfolk Island before branching to Southport, Australia, and Doubtless Bay, New Zealand—remained exclusively on imperial territory, even though there were a number of alternatives that offered shorter undersea segments and less expensive installations.[26]

These links were insulated by empire, but this meant that in some places the network lacked sufficient interconnections with local infrastructures. In 1911 the Union Steam Ship Company terminated service to Fanning Island. This threatened the Pacific Cable Board's ability to maintain its cable station, and it began to depend on ships servicing its American competitor, the Commercial Cable Company, and a transportation route through Hawai'i, an island it had avoided landing the cable on for the sake of maintaining the network's imperial cohesion. The New Zealand General Assembly observed that this was quite an "anomalous position," in which the company depended "on the benevolence of a non-British company for maintaining communication with a station which was chosen, in spite of many disadvantages, for the single reason that it fulfilled the condition of being British territory."[27] Even if the network's maintenance and support systems relied on interconnections with foreign networks, they were still seen as secure so long as they remained on imperial territory.

Parallel to the Canadian impetus for the first north-south transpacific link, in the context of growing imperial rivalry, Americans called for a transpacific cable as a "national necessity," initially conceptualizing it as a state project, with "manufacture, laying and operating . . . 'wholly under the control of the United States.'"[28] Similar to the Pacific Cable Board, the U.S. Commercial Pacific cable—which extended from San Francisco to Hawai'i, Midway Island, and Guam before branching to Japan and the Philippines—was not intended to connect sites that were off the grid. Instead, the cable was to help the United States drive demand and exert "commercial and political influence" in the Pacific and was motivated in part by the entrepreneur John Mackay's quest to establish cheaper transpacific communication.[29] In order to interconnect its cable with Asian networks, however, the Commercial Pacific Cable Company ultimately had to sacrifice three-quarters of its ownership to non-American companies.[30] But as Winseck and Pike note, the company "had no problem claiming to be 'as American as apple pie' when standing before Congress to promote why it should be chosen to lay the first U.S. cable across the Pacific."[31] Again similar to the Canadian cable, the American cable's security was ensured by the extension of the cable through U.S. territory, if not by national ownership of the company.

The transpacific cables, charting new and unprecedented routes across the Pacific, were not easy, quick, or lucrative projects. They mastered the scale of the globe: the Pacific Cable Board line between Bamfield, Canada, and Fanning Island would be the longest undersea cable in the world. They also required new expenditures in infrastructure and the production of environmental knowledge. A new 8,000-ton ship, the *Colonia*, had to be constructed specifically for the Pacific Cable Board line. Surveys of unknown parts of the Pacific were conducted. These efforts were exerted in the interest of interconnecting the new cable with existing networks to generate competition and break monopolies up, and they were motivated by arguments to provide a more secure network via diverse undersea (rather than terrestrial) routes. Conceptualized as the extension of multiple cables to the same endpoint (rather than multiple installations at the endpoints), diversity of routing became a key strategy of insulation for telegraph traffic.[32]

Despite the fact that dominant discourses during this period portrayed water as a threat to telegraphic communication, the resource-intensive, time-consuming, and less economically feasible approach of laying undersea cables was chosen because in many cases it appeared to be the more secure option. However, the approach to securing cables shifted as perceived threats changed: companies moved from a concern about protection from the ocean to a preference for using the ocean's depths as a layer of protection from potential colonial unrest, rival nations, and ships' anchors. As a result, many of the global networks of the late nineteenth century were routed underwater rather than over land.[33] In this context, the transpacific cables, justified on the basis of diversity, positioned Canada and the United States as key centers for traffic and set the stage for the geography of subsequent Pacific communications. Once a secure route had been established, the conservative cable industry was reluctant to forge new paths. Although the expenditures eventually became far greater than the strategic payoff, and the imperial logic and strategic nationalism governing these systems waned, the colonial telegraph network established a set of contours that would affect cabling throughout the next century.[34]

The Coaxial Cold War

In the 1950s analogue coaxial cables were laid under the oceans. The geographies of these new systems evidenced a negotiation between historical contours and emerging forces, including the reconceptualization of security during the Cold War, the nationalization of telecommunications, and the evolution of a club system of construction. Coaxial cables had been first developed in the 1930s, but at that time they could reach only short distances because their signals attenuated after about a hundred miles. In the 1940s the invention of the submersible repeater, a technology that amplified the cable's signal on the bottom of the seafloor, made transoceanic speech transmission possible. Perhaps to forecast the coming of a new era, the first deepwater cable with repeaters was American, laid in 1950 between Florida and Cuba. In 1956 Transatlantic No. 1 (TAT-1) linked Scotland and Newfoundland. These new systems were able to carry not only telephone conversations, but also telegraph messages, telex transmissions, photo telegrams, and even slow-scan television. They required a major shift in technological production: cable ships had to be modified or constructed anew to handle the repeaters; new cable stations had to be created to house terminal equipment; and plastic polythene insulation replaced gutta-percha, which meant that cable companies no longer needed to extract that material from Southeast Asia. This provided an opportunity for the companies to develop both technologically and geographically, and the culture of the Cold War affected their decisions about the new networks' topography.

One of the primary justifications for laying coaxial systems in this period was their perceived security. Undersea cables had been framed as a secure technology after World War I and the later expansion of shortwave radio in the 1920s and 1930s. In comparison to radio, cables were much less vulnerable to interception and remained unaffected by atmospheric conditions and sunspot activity. Therefore, rather than being replaced by radio, cables were deemed a necessary supplement to it. As the naval correspondent of the *Morning Post* reported in the 1920s, "The truth of the matter is that wireless and cable communications supplement one another . . . but the use of the cable, provided that its terminals and intermediate stations are situated on British soil, has one outstanding advantage over wireless—secrecy."[35] Indeed, when the Imperial Wireless and Cable Conference merged wireless and wired networks (forcing the Marconi and Eastern Telegraph Companies together) in 1928, a key argument for keeping the cable networks rather than moving exclusively to wireless was this perceived security. With the advent of wireless, network diversity no longer entailed laying two cables but required the use of multiple transmission technologies. As Herbert Schenck wrote in 1968, "World-wide defense requires world-wide telecommunications. Prudence suggests that these be not concentrated in any one particular mode of transmission."[36] When satellite communication was developed, this only enhanced the perception of cables as secure: they offered insulation for signals that might be easily dispersed or intercepted via the airwaves.

The U.S. military needed secure communications before and during the Cold War, and coaxial cables were seen as giving stability, and presumably secrecy, to international calls. As James Schwoch has observed, "From an American perspective, the interplay of extraterritorialities with global television and electronic information networks was first conceptualized as a security issue. Arguments, positions, rhetoric, and discourse invariably began with an articulation of security concerns."[37] An engineer working during this period recollects that the "security issue" always formed a backdrop for cable projects: "submarine cables always were regarded more secure than satellite circuits. . . . It wasn't spoken about too much, but it was acknowledged."[38] Along many routes, transoceanic telegraph cables were not disconnected until coaxial networks were implemented. Bernard Finn reports that it was well into the satellite era by the time the last significant telegraph cables were removed from service in 1973.[39] Just as coaxial cables did not replace radio (and in many cases additional radio circuits were used to back the cable systems up), neither did satellites later replace the cables. The need for secure routing and diversity via multiple transmission technologies ensured cable's continued presence throughout these periods.[40]

The new geographies of the coaxial network, especially in the United States, reflected the logics of the Cold War and accorded with a dominant strategy of infrastructural decentralization that located stations in remote sites so that they would not be as vulnerable to either nuclear attack or marine traffic.[41] At the end of the telegraph period there had been a general centralization of cable networks around the southern Pacific—a movement reflecting the conceptions of security leading up to and following World War I (cables were perceived as being more secure if they were near urban hubs that would facilitate military protection) as well as transitions in cable technology that made fully staffed cable stations less important.[42] Remote cable stations generally moved closer to cities, where it would be cheaper to hook into existing infrastructure, but they were often located outside the urban hubs.[43] This reorganization set the stage for the coaxial installations: cables could not be in places where there was no infrastructure, so they could not be too far from urban hubs; nor could they be so close that they would be disrupted by other forms of traffic. In the telegraph network, San Francisco had been the Pacific's gateway to the United States; coaxial cable landings were relocated to Manchester (a rural coastal town north of San Francisco) and San Luis Obispo (midway between Los Angeles and San Francisco). In Hawai'i, cables no longer landed centrally in Honolulu but connected O'ahu's less populated west and east coasts.[44] Similar transitions happened in Asia and the South Pacific, as stations at Baler (in the Philippines), Ninomiya (in Japan), and Lauthala Bay (in Fiji) were dislocated from the countries' urban centers. This strategy also served to protect cables from turbulent urban circulations, including nautical traffic that threatened to sever cables, costs of coastal property that made construction and maintenance more difficult, and swelling populations that could inadvertently disrupt cable traffic. Moreover, although in the colonial period, securing one's network entailed routing two cables to the same station, in the Cold War era companies often saw it as establishing a range of diverse landing points and stations.

Telecommunications companies had to balance the drive to decentralize with the traction of existing cable installations. In some places, if failures had not occurred there seemed little reason to relocate. When determining where the Australian terminal station would be located, the management committee for the 1963 Commonwealth Pacific Cable (COMPAC) considered several new sites, including suburban and rural locales. The committee members noted that, given the "premium on space in the city centre," it might be preferable to establish their new station in one of the suburbs.[45] In the end, they decided to route the cable into the existing city hub, in part because of the difficulty of developing a new route: that would entail digging up roads and demolishing buildings to connect the old network with a new landing. When landing facilities, a cable station, and a conduit from the beach already exist, as one engineer observed, "the costs of diversity are quite high."[46] Even if a less central location would have been preferred, the existing gateway had traction: it was appropriately insulated from and connected with the surrounding geography.

Despite the general shift toward decentralized landing points, coaxial cable routes generally followed the same paths as the two original transpacific telegraph cables. Terrestrial routes were avoided even when that meant a longer marine segment: the ocean continued to be used as a means of insulation.[47] The first transpacific coaxial cable, Trans-Pacific Cable 1 (TPC-1), followed the general path of the Commercial Cable Company's cable, moving from California to O'ahu, Midway Island, and Guam before linking to Japan and the Philippines and building on the existing economic, cultural, and technological interconnections with these locations. COMPAC followed the original Pacific Cable Board route, since that had a "low incidence of faults."[48] The builders argued that at "Fanning Island and Fiji the records of the telegraph cable landings had proved the reliability of the approaches and landings and there was no reason to consider alternatives."[49] This became the industry's standard practice, as the ROGUCCI report directly states: "If an existing route is known to be acceptably reliable from natural or man-made damage then there is a great commercial incentive to lay cables along the same route."[50] Like the earlier transpacific links, many of these cables were laid between countries that already had international voice communication via radio, with the intention of providing diversity of routing to established markets and users.

With the breakup of colonial empires, the focus of securing the cable network shifted from routing via one's territory or colonial holdings to having national control over the processes of building, operating, and maintaining the cable network. Although the British continued to control many of these resources, new players such as the United States and Japan emerged. In the United States AT&T became a major investor in and developer of undersea cable systems. The company created a specialized submarine systems laying group, AT&T-SSI, which had its own fleet of cable ships. Simplex, an American company that had been involved in cable manufacturing since the nineteenth century, invested in the construction of a new facility in 1953 and helped build the American portion of the first transatlantic telephone cable. The U.S. Underseas Cable Corporation was created in 1959 to manage cable projects from inception to completion. These companies also helped facilitate the U.S. military expansion in the Pacific. AT&T's transpacific cable provided circuits for the U.S. Defense Communications System during the Vietnam War. Even though microwave radio links were used for command and control, to obtain reliability "through diversity of route and mode," the military turned to cables.[51] The strengthening of U.S. cable suppliers also facilitated the extension of military-only networks, including both surveillance cables and links between South Vietnam and the Philippines and between Johnston Island and O'ahu, as well as a coastal cable project in Vietnam. As in the case of the All-Red Line, the routing and geographies of these links were often influenced by strategic investments.[52]

Although the Japanese had long been invested in regional telegraph systems, in the coaxial era they too developed a larger-scale network. The Kokusai Denshin Denwa Company (KDD) was formed by the Japanese parliament in 1953 as a monopoly international telecommunications carrier, partnering with the Americans on transpacific and East Asian cable projects.[53] The company subsequently introduced new coaxial cable technologies, invented a new method of burying cables, and built its own cable ship. Nippon Electric Company assisted in cable development, and Ocean Cable Company of Japan supplied cable for a number of these projects. One telecommunications historian argues that in this period "Japan's potentialities to plan, manufacture, construct, maintain, and operate coaxial submarine cable systems [had] been developed to reach the highest technical standard of the world."[54] The national development of a cable industry was a strategy to insulate cables from the potential interference of other nations in their cable construction and operation, since everything could be controlled and accounted for by the incumbent telecommunications company.

During this period, a new model of institutional interconnection, the club or consortium system, displaced the imperial model and eventually introduced a new set of concerns to cable routing. In the club system, cables were no longer designed and operated by a single company (and insulated by empire). Instead, cable construction, engineering, and route design became topics of negotiation on cable committees whose members represented nationally owned or affiliated monopoly telecommunications carriers, whether American, Japanese, British, or French.[55] Committee members first agreed to jointly build a new network. Each company would contribute its expertise and take responsibility for facilitating interactions with its government, developing landing arrangements, and negotiating domestic connections. Although telegraph cables had often been planned and operated by one company, with coaxial systems the duties of maintenance, the costs of construction, and the revenues generated were typically shared more broadly. For example, for COMPAC the participating countries agreed that the cost of £26.3 million would be split among Canada (£8.8 million), the United Kingdom (£8.3 million), Australia (£6.6 million), and New Zealand (£2.6 million).[56] As Richard Collins argues, this evidenced a "supercession of imperial organisational arrangements by national control," one in which self-contained networks moved toward "an integrated global system."[57] "Cooperation was the name of the game in those days," notes Jean Devos, of France's Alcatel Submarine Networks. Through the club structure, "a sense of fraternity and mutual respect developed over the years between all the players."[58]

The terrain to be secured was no longer the country's territory, but the circuits of the cable: each nationally affiliated company would own a percentage of the cable or a specific number of circuits. Typically, companies owned out to the midpoint of a cable (termed a *half circuit*), and the capacity would be allocated before the cable started operation. This structure limited the development of a market for international traffic, since anyone who wanted to provide services using the cable would have to negotiate with multiple owners rather than just one. For example, in the case of the COMPAC cable, AT&T "vigorously" pursued an offer to purchase a 25 percent share in the cable south of Hawai'i, but the "Management Committee considered this proposal ran contrary to the concept of the Commonwealth cable system . . . its acceptance would compromise the Commonwealth nature of the venture and reduce flexibility in control and development."[59] Anyone not in the club could easily be kept out.

Operators would pool their resources to benefit all of the club's participants, yet their interconnection depended on being insulated from competition. The monopolization of international communications by only a few players and the domination of construction by a handful of companies (a club cannot be too big) were critical to this new system. Most often, even when a large number of parties bought into the system, the design and construction of the undersea segment would be handled by one of the large companies.[60] Any country needing access to international networks would have to choose between these suppliers "according to natural areas of political, cultural, economic and financial influence."[61] One engineer described the process this way: "In putting a bid together, if it was for a Cable & Wireless project, I would be foolish to use anybody but Cable & Wireless for our marine installer, because I wouldn't get the work. It was recirculatory money."[62] These trusted alignments and partnerships protected the club's shared investments. As another engineer recalled, club members would be unlikely to work in an environment where "politically you're not familiar with how they do things. There was the security aspect, which was never mentioned too loudly, but it was always in the background, you didn't want to put your cables at a risk. There was not a name for what sort of thing could happen."[63] In brokering interconnections between nations, the club made it more secure for each country to extend circuits into foreign territories.

The result of this new institutional arrangement meant that, even though the practices of cable maintenance and manufacture remained guarded by individual nations, there were now more technological connections between the systems. The club system introduced extensive interconnection agreements to the planning process. Whereas in the colonial era, the British-affiliated cable steered clear of Hawai'i in favor of Fanning Island, under the club system the planners observed that even though Hawai'i was U.S. territory, "a landing there would have substantial advantages from the technical point of view. Moreover, it would provide a junction between the Commonwealth and the American telephone cable systems and yield material financial advantages."[64] This interconnection would also afford diversity of routing, offering a possible backup in case of cable failure, and would be cheaper to maintain.[65] As a result, even though the Commonwealth system would not sell shares to the

United States, it agreed that "there was every advantage in ensuring that the Commonwealth plans and the United States plans were complementary and, as far as practicable, coordinated."[66] In return for the connection at Hawai'i, the U.S. Defense Department received dedicated backup circuits to Canada as well as circuits through Australia.[67] The land on which the station was placed was reportedly given to the Commonwealth in exchange for land in Australia that could house an American military base. However, the landing country retained oversight: in the United States, the Federal Communications Commission stated that the Commonwealth cable must conform "with plans approved by the secretary of the Army" and that it should be shifted (at the expense of the Commonwealth) if the secretary of the Army ever required that.[68]

The geographies of coaxial cables reflected existing route lines, seen as secure because they hadn't had many disruptions, and followed the original investments of telegraph companies. However, companies also had to negotiate the emerging cultural conditions of the Cold War era. Although cables continued to be viewed in terms of security, the companies' understanding of security itself changed, shifting to encompass strategies such as infrastructural decentralization. The nationalization of telecommunications meant that the territory of empire could no longer serve as a layer of insulation. As a result, via the club system, a new set of connections—made at sites such as Hawai'i and Guam—were understood to make the cable network more secure. This enabled countries to extend circuits into other nations without encountering interference (which would be mitigated by the local partner). In addition, by linking to each others' networks, national carriers could achieve a network of circuits with physical redundancy: the TPC systems across the northern Pacific and the COMPAC and South-East Asia Commonwealth Cable (SEACOM) systems across the southern Pacific, even though "owned" out to mid-Pacific by nations on either side, were configured as a loop. Individual carriers (as long as they were part of the club) were able to buy capacity in an entire ring. Routing signals the long way around this ring would still be more reliable than wireless. As in the telegraph period, these transpacific links, made possible by historically specific geographic negotiations, were not always commercially profitable at their inception, did not always connect previously unconnected sites, and did not simply follow the shortest paths between cities. Geoff Parr, a global manager at Australia's largest telecommunications company, recollected that "it didn't really matter if you went a circuitous route," as long as you got to where you were going.[69]

Fiber-Optic Financialization

From the late 1980s to early 1990s, two major shifts—the development of fiber-optic technologies and the process of telecommunications deregulation—significantly changed the culture of cable laying and, as a result, the network's geography. However, the first fiber-optic cables continued to follow the contours of telegraph and coaxial networks, giving the imagined sources of friction and security of those periods—from the debates over territorial security in the colonial era to the spatial decentralization and institutional interconnections of the Cold War era—a residual life in digital communications systems. Fiber-optic systems, which encoded signals in light and sent them down a thread of glass, had been in development since the early 1980s, but given the conservatism of the industry (in which nothing is adopted until it is tried and true), the first international system was not put in place until 1986. Even with this technological advance, the first fiber networks were built according to the consortium model: they were based on standard traffic forecasts, intended to simply increase the capacity available to partner countries, and laid by companies with a long history of cable experience. They reinforced the nodes of the existing telecommunications networks. In 1989 AT&T laid the first transpacific fiber-optic cable, Trans-Pacific Cable 3 (TPC-3), between Cold War–era stations at Makaha, O'ahu; Tanguisson Point, Guam; and Chikura, Japan.[70] PacRim (1993), the first fiber system in the South Pacific, was routed almost exactly along the path of the previous analogue cable, Australia–New Zealand–Canada (ANZCAN) (1984), which had in turn followed the COMPAC route. Terrestrial segments continued to be most dangerous part of the system, to be avoided whenever possible.

As in the coaxial era, telecommunications companies would occasionally consider new paths but ultimately determined that existing routes were more secure. For example, the telecommunications company laying the Australia-Japan Cable in 2001 considered alternative routes to Sydney but determined that they were too difficult. Failed attempts to pioneer new routes demonstrated these risks. When laying JASURAUS in 1997 from Jakarta to Australia, Telstra decided to develop a new landing in Port Hedland, Australia, to connect with a domestic fiber network. To do so, engineers had to design an elaborate trenching scheme to bury the cable fifty kilometers out to sea along a long shallow shelf. After this drawn out and exasperating experience, the company returned to the previous west coast landing in Perth for Southeast Asia–Middle East–Western Europe 3 (SEA-ME-WE 3) in 2000. When routes were altered, this was typically justified in terms of the need to secure a cable from nodes and landing points seen as problematic and to connect it to other networks, including satellite systems.[71] The PacRim system skipped Fiji, which had had several coups, and Norfolk Island, where cables were notoriously difficult to install.

As fiber-optic technologies improved—and data communications networks grew with them—they greatly increased capacity from earlier links: the amount of traffic that new systems could carry increased dramatically every year, and this process helped speed up network development. Although PacRim East had two fiber pairs of 560 megabits per second in 1993 (one telecommunications manager involved in the project recalled wondering "who would want any more than that?"), a mere three years later JASURAUS was completed with five times as much capacity.[72] Throughout, the cost to build systems stayed relatively constant. As a result, many of these early fiber-optic cables reached economic obsolescence before technological obsolescence. The PacRim network was designed to meet demand through 2005 but was full by the end of 1996.[73] The first generation of fiber-optic systems had some of the shortest life spans in cable history. Although the first transpacific telegraph cable lasted over half a century, and the first coaxial cable in the Pacific was in service for thirty-two years, the first fiber link to Hawai'i lasted only fifteen. Builders asked whether there was still a need to engineer to the same set of standards: did they really need a twenty-five-year design life, which put a lot of cost into development, when systems were becoming obsolete within a decade? Although these questions subsided in the late 1990s when new technologies made it possible to transmit more information over existing cables, they had the temporary effect of accelerating network development (helping to generate a cable-laying boom) and causing the industry to question the longevity, and indeed some of the historical foundations, of their systems.[74]

At the same time, deregulation introduced a new alternative to the club system. By the 1980s and 1990s, interconnection agreements and joint ownership had become quite unwieldy. Fourteen companies eventually signed the construction and maintenance agreement for ANZCAN: countries from Papua New Guinea to the Philippines and Fiji were allowed to invest. TPC-3 had seventeen owners, Guam–Philippines–Taiwan (GPT) had twelve, and Hong Kong–Japan–Korea (HJK) had thirteen. As state-owned monopolies were privatized and international telecommunications became subject to competition, the opening up of undersea cable planning intensified. In 1984 the monopoly of AT&T, the world's largest telecommunications provider, was broken up. The Japanese began to deregulate the monopoly of the Nippon Telegraph and Telephone Corporation in 1985.[75] British Telecom and France Telecom were sold to private investors. This created opportunities for both smaller carriers and larger companies to expand and operate international telecommunications in places previously protected by the club's interests.

With the expansion of privatization, many cable projects became speculative money-making ventures that involved calculating market risk, rather than endeavors to meet forecasted demand. Companies that had been "supporters of national industry and public service" became "business-oriented organizations" and moved away from the "open collaboration" and fraternity with fellow companies.[76] The Internet boom made it increasingly difficult to base financing on a traffic forecast, which was "becoming a crystal ball exercise."[77] One entrepreneur in the cable industry remembers his early encounters with the culture of consortium cable work: the consortium had meetings that would go on for days, "almost like formal diplomacy," with highly technical discussions that helped establish the participants' credibility.[78] Increasingly, however, engineers had to articulate and design the system in relation to potential financial gains, and a new group of people in marketing and sales became involved in discussions about cable routing. Without the club's coordination, individual companies took less responsibility for planning a robust global network.

Together, these forces—the quickening pace of cable development, the transitions of deregulation and privatization, and the entry of new players—led to a set of efforts to develop new cable systems that deviated from the club model. The first private cable was the transatlantic cable PTAT-1, which was completed in 1989. Andrew Lipman, the lawyer for the system, recalls that the private model had been pioneered already in satellites, and to some extent, the challenge was a psychological one: the entrepreneurs behind the system had to convince the cable industry that it could be done. Their success sparked a rash of other private cables. The formula spread to the Pacific when North Pacific Cable (NPC) was established to compete with a consortium cable, TPC-3. Developed by companies that were historically foreign to the transpacific cable business (and increasingly including nontelecommunications companies), these newer cables—in part because of the dynamics of capacity—often cost less and carried more than earlier networks, and thus undercut their rates. By 1997 the $1.2 billion Fiber-Optic Link Around the Globe (FLAG) network, backed by over thirty financial institutions, had been laid along a route between Europe and Asia traditionally managed by a consortium. This cable initiated the sponsor's approach, in which a private company organized a project and allowed others—not limited to companies who had financed the network—to buy capacity on it as they needed.[79]

Overall, this process made the distribution of circuits more flexible. The terrain to be guarded in the consortium era—ownership of a circuit—was not the site of contestation. No longer held up by the diverging opinions of numerous committee members, the timetable for private cable projects became shorter to capitalize on the current market: instead of taking five years or longer, cables could be launched and completed in under two years. There was an urgency to this development, since in some cases, if the system was operating by a certain date, precommitted customers would be able to back out.[80] In combination with the new technology, this led to a more risky business environment, although some of these risks shifted from major telecommunications carriers to independent companies and financial institutions.

Deregulation, privatization, speculation about increases in capacity that would be needed by the growing Internet (to be stimulated by high-bandwidth–intensive applications such as videoconferencing), and assumptions about an inherent relationship between telecommunications investment and economic development together drove a global boom in cable laying. Stewart Ash, who worked in cable installation at the time, recollected that "cable manufacturers were building factories all over the place. We had a cable factory in Southampton in the U.K. We had one in Portland, Oregon. We had another one in Australia. People were pushing out 100,000 kilometers of cable a year."[81] The geographic expansion of undersea cables increased at an unprecedented rate.[82] New cable ships were built. The strict national affiliation with cable suppliers was severed, as many companies strove for the best deal and could choose among suppliers from various countries.[83] The invention of the universal joint in the 1980s, an undersea cable technology that enabled the splicing together of different kinds of cables from different companies, further broke down the segmentation of marine services across nations. These shifts, like much of the globalization of the 1990s, shook up—though they did not eliminate—the national entrenchment of postwar cable development. New companies such as Global Crossing, 360Networks, FLAG, Tyco, Level 3, and Singtel/C2C—many of which were based in the United States and had little experience in the industry—entered the cable business and began to compete with older telecommunications companies.[84]

This transition had several significant impacts on the geography of undersea networks. First, the historically complementary relationship between wireless and wired transmissions that had been in place since the development of wireless telegraphy was fundamentally altered. Satellites ceased to be a sufficient backup, and the financial impacts of cable breaks increased.[85] When there was a major outage on PacRim West in the mid-1990s, the network did not have enough satellite capacity to restore its traffic. Brett O'Riley, a telecommunications policy expert in New Zealand, recalls that this was "a wake-up call to the industry" about the imbalance between wired and wireless capacity.[86] This transition was a fundamental shift in network geography that brought the industry back to the telegraph era: cables once again needed cable backups for diversity. Given the increases in how much each system could carry, the content carried by a single new cable might exceed that in all prior systems. Even old cables could not provide backup for the newest links. As a result, companies began to develop ring systems, more extensive than any before, that connected a series of locations so that there were essentially two paths between any given points. These systems embedded diversity in the geography of each cable system and further insulated traffic from any potential instability in individual locations. In 1996 the consortium cable Trans-Pacific Cable 5 (TPC-5) pioneered a ring structure around the Pacific, with two points in Japan and four in the United States, providing instantaneous backup if either leg went down. Following this, four other ring networks, each of which cost over a billion dollars to construct, were built with northern and southern transpacific routes.[87]

Combined with the loss of wireless backups, the technological development of the underwater branching unit, which enabled companies to redirect signals at the bottom of the ocean, pushed the cable industry to develop new transpacific geographies. Previously, to connect any three locations one would have to use a ring structure or choose one location in the network as a central point. With branching units, the cable station was effectively moved underwater: the strategy of insulation—using the ocean to protect communication—developed in the colonial era now enveloped nodes in the network itself. The development of trunk and branch systems made routes more diverse and resilient, since they could not be cut off by a failure at a station in the middle. They also required more planning in the beginning, since the system could not easily be expanded once it had been laid. The Australia-Japan Cable introduced a new mode of diversifying landing points in the Pacific: at each of its landings in Australia, Guam, and Japan, the cable branched into two sections. Since the most dangerous areas of the network were the shore ends, this part—rather than the secure deepwater segments—was duplicated. This network geography inverted the approach of the telegraph era, in which multiple deep-sea cables were laid to the same endpoint; instead, security entailed using one cable with multiple endpoints. At the same time, a collapsed ring structure was developed: following the trunk and branch design, this added diverse technological pathways between points, though they followed the same physical route (figure 1.3). These investments in diversity were conducted at the level of the individual network, without necessarily taking into consideration the stability of the system as a whole.

FIGURE 1.3. Changes in network design.

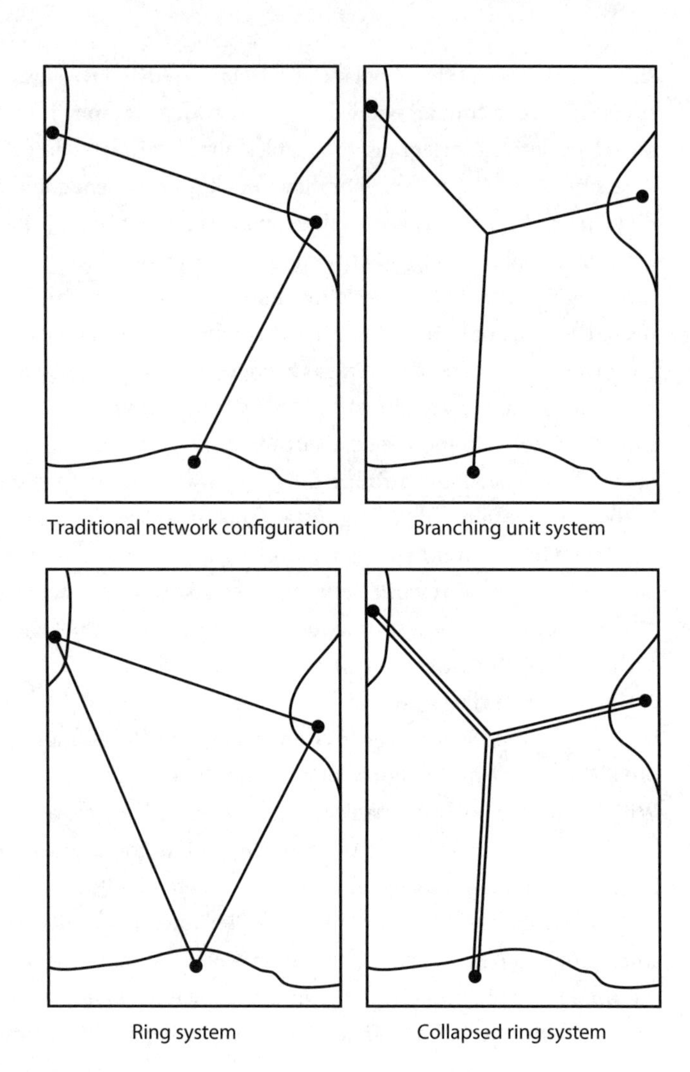

Private cables, run by independent companies, also developed a new geography of landing points, since existing telecommunications companies in many places blocked the new networks from terminating at their stations. North Pacific Cable built a new cable station at Pacific City, Oregon, pioneering the landing of cables on the northern Oregon coast. Global Crossing established a new station at Grover Beach, California, for its transpacific and South American cables. Many historical monopolies resisted the change. However, in the end, the new nodes were connected with the old ones, and the boom helped create duplicate routes across the oceans and, ultimately, a more secure network.[88] The drive to diversify and back-up one's networks was strong:

established companies acquired capacity on the new cables to provide backup routes, and in many places new cable companies purchased backup capacity on the older cables. The club system did not disappear, since the historically dominant and nationally affiliated telecommunications companies laid new consortium cables to compete with newcomers.[89]

The Southern Cross Cable Network (SCCN) is exemplary of the era's shifts. By the late 1990s, the PacRim system, the only cable network out of Australia and New Zealand, had filled up. One telecommunications worker claimed that the PacRim consortium "was manipulated by Telstra and AT&T deliberately to keep other players out."[90] He recalled a meeting in Queenstown when representatives of a number of telecommunications companies showed up to discuss how the remaining capacity could be used, only to find that Telstra and AT&T had made a deal before the meeting, allocating all capacity and blocking new players' access to the network. This "created a huge amount of bitterness in the industry," and as a result, a group of New Zealand telecommunications workers began to think about building their own network. Like many of the private systems, SCCN, even though partially owned by the nationally affiliated Telecom New Zealand, was driven by companies that had been shut out of a consortium but needed capacity to participate in the 1990s technology boom.

Simply building a cable to Australia was not a prospect for economic reasons. Even if New Zealand companies could send data to Australia, they would still face the same problem of depending on the PacRim cable (and the former national monopoly, Telstra) to reach the rest of the world. Since north-south links across the Pacific were difficult and expensive to build, New Zealand could not finance a transpacific link based on its users alone, and it could attract little interest from North American, Asian, and European companies. As the company did not have backing or up-front financing from existing telecommunications carriers, it had to devise an economic plan that would convince financial institutions to lend it hundreds of millions of dollars. One of the original cable builders, Charles Jarvie, recalled: "There's got to be a story for the bank. They won't just take your word. They'll say, 'Well, who's giving credit support to this contract? Who's standing behind you?'"[91] No matter how enthusiastic the builders were about the social or economic benefit for New Zealand's industries and emerging Internet users, they were forced to develop a commercially plausible rationale for the system.

Brett O'Riley, who helped to develop the cable network, remembers that "Southern Cross was different, because for the first time we really treated it as a market exercise. This was not industry getting together and saying, 'How do we make this work?' This was sponsors . . . we had a marketing committee. We designed a logo for it. We designed merchandise for it."[92] Even SCCN's name speaks to the shift in cable politics: the Southern Cross is the dominant constellation in the southern hemisphere. The SCCN team wanted a name that was more interesting than the typical engineer-produced name (such as TPC-3) and that would signify that it was a different kind of cable system. Fiona Beck, SCCN's CEO, had a background not in engineering but in finance and strategy, which helped make the cable company aware of the kinds of market-related questions that would be asked of them.

As a result, SCCN's developers conceptualized the network as a transpacific link from the United States to Australia, positioning New Zealand as a stopping point—in effect, using Australia to provide for New Zealand. As one of the builders joked, "The New Zealand way is to say, 'These Australians won't think about us,' that 'we'll have to convince them it's the best idea they've ever had.'"[93] Like many private cable system developers, they sought to move quickly and position SCCN as a more dynamic company, free from the large committees and extended timetable of the "politically based" consortium cables.[94] The network was designed as an Internet-based cable that would be the fastest and most straightforward route to the United States, where most of the Internet servers were. Inspired by FLAG and other independent projects, the Southern Cross team established partnerships with companies in Australia and the United States that had also been left out of major cable consortia. They were ultimately successful in getting funding (the network cost $1.5 billion), and the system went live on November 15, 2000. As using its own transpacific cables became less feasible—PacRim became economically obsolete and the Australia-Japan Cable proved to be a truly circuitous route to the United States—Telstra, which had initially blocked New Zealand's attempts to get capacity, ended up becoming one of Southern Cross's largest customers.[95]

Because of its oppositional relationship with the major telecommunications companies, SCCN established a new geography of cable landings that included setting up new landing zones, stations, and local agreements. New stations were constructed in Australia at Alexandria and Brookvale (separate from, but close enough to Telstra's Sydney stations), in Hawai'i at Kahe Point (just down the shore from AT&T's station) and in California at Morro Bay (near the existing AT&T station at San Luis Obispo). The Southern Cross stations were open access, and new companies were allowed to connect their cables with them. Although PacRim had skipped Fiji because it was a problematic node, Southern Cross stopped there, accepting the Fijian telecommunications company as a partner in the project. Landings at New Zealand and Fiji, because the partner organizations there were descendants of Cold War–era public telecommunications companies, were in the sites that had been established in the 1960s. In addition, SCCN was planned as an elaborate ring system, another form of differentiation that gave customers diversity and protection. The network's geography—like that of many networks of the period—evidenced a compromise between the need to connect with existing nodes and the spatial pulls of the new cultural environment.

Just as SCCN was being put in the water, the cable market collapsed as a result of the excess capacity offered by the numerous cable systems laid in the late 1990s. The timing of Southern Cross was perfect, as its builders recollect, poised on the crest of the cable boom. Barney Warf explains that, due to the cable boom, between 1988 and 2003 the number of voice paths across the Pacific increased from 1,800 to 1.87 billion, and at the end of this period, fiber-optic cables had 94.4 percent of the world's transmission capacity (compared to only 16 percent in 1988).[96] Corresponding to this, the price of transmission dropped drastically, sometimes by as much as 90 percent.[97] One cable engineer explains:

> Banks decided you could have as much money as you want, as long as you wrote in the back of your cigarette packet "dot com" or "submarine cable." They would say, "How many millions do you want?" You had this stupid situation where five or six people wanted to build a cable across the Atlantic, and they all guaranteed that they were going to get 100 percent of the traffic that would be going on those cables. Of course, what happened was that when the five were built, they all got 20 percent each and they all went bust.[98]

Many companies could not recoup the hundreds of millions of dollars expended to lay these networks.[99] 360Networks, Global Crossing, FLAG, Level 3, and WorldCom all went bankrupt between 2001 and 2002; some were acquired for cents on the dollar. In total, the market fell from over $10 billion per year to only a few hundred million after 2002, and control shifted to South Asian operators: some of the largest owners of undersea cable networks are now in India (Reliance Globalcom and Tata Communications).[100]

Through the cable bust and the financial cuts of privatization, companies lost a significant part of their labor force, and with it, these workers' knowledge and experience. By 2001 Alcatel had reduced its workforce in undersea networks by 48 percent and closed its U.S. plant in Portland and an Australian one in Port Botany.[101] In the period 1991–95, AT&T cut 123,000 jobs, 30 percent of its labor force.[102] Years later, as the industry began to rebuild itself, knowledge about the way to do many things had been lost. Stewart Ash described the situation: "Over the last 160 years a lot of mistakes have been [made], and people have learned from them. The best companies have secured that knowledge in written documentation, procedures, etc. Others have maintained the knowledge in the heads of the experts that they had in the organization. In the bust, a lot of that experience walked out the door."[103] He observed that with its regeneration since around 2005, the cable industry now lacks experience. "History is valuable because the experience informs the way forward," he says. "Mistakes made before shouldn't be made again. Unfortunately, if you tend to lose expertise, the same mistakes repeat themselves." After the bust, some things have gone back to normal: the consortium system made a significant comeback, and there are even some hybrids that have aspects of both consortium and private cables. However, people in the cable industry fear that failing to learn from the past will result in a less stable and secure network in the long run.

Diversity: Three Guys and a PowerPoint

At the turn of the twenty-first century, a number of frictional forces—along with the cable bust—challenged the industry, tempering its exaggerated claims to easily move through time and space. Nautical traffic was reaching new heights, with the intensification of fishing and shipping in coastal waters. Cables were regularly endangered by ships at pressure points such as the Strait of Malacca, between Indonesia and Malaysia. Although in the consortium model national boundaries were rarely crossed by foreign companies, which typically "respected each other's sovereignty" and let each national partner negotiate its own landings with its respective national authorities, with the new geography charted by private companies, a single company often had to negotiate landings in many nations where they were unfamiliar with established practices.[104] Conflicts occurred at landing zones as the multitude of private cables, whose entry was no longer mediated by a nationally affiliated company, crossed paths and overlaid one another. Although it had been common practice to lay new systems along existing routes, especially in areas that had a low level of faults, in the era of quickly paced competition, companies extended cables into more risky areas and put less preparation into securing routes, which led to an increase in faults.

Following the cable bust, a set of high-profile breaks drew attention to the limitations of existing routes. In December 2006 the Hengchun earthquake off the coast of Taiwan, a key destination for Asian traffic, triggered underwater landslides that caused nine breaks in seven different cables, significantly decreasing Internet connectivity and the quality of phone calls.[105] The repairs took seven weeks and required the use of cable ships from around the world.[106] This stopped Internet traffic, which in turn prompted articles in the media and brought a broader recognition of the significance of undersea networks. One cable station manager recalls this as the point when Internet cables first became visible to the public, since "people don't tend to worry about things until they're personally affected."[107] In 2008 a series of cable cuts occurred just north of Alexandria, Egypt, a pressure point in Europe-Asia traffic. Haiti's 2010 earthquake disrupted a cable station and the country's underwater connection. The following year, the Japanese tsunami broke several cables. Repairing such systems quickly can be difficult given the scarcity and wide dispersion of capable repair ships, which are allocated to specific zones of coverage and not always located close to a problem. Cable owners might be able to call in a ship from another zone (especially in the event of several breaks), but at times they have to wait for weeks until a ship is available.[108] Additional delays can be caused by weather or the need to acquire permits before making repairs in some nations' waters. After the Japanese tsunami, cable repair ships would not go in local waters for fear of radiation contamination.

Concerns about deliberate attacks and a new "asymmetric terrorist threat" escalated in the era after 9/11, especially given the widespread dependence on these systems.[109] Cables are now considered *critical infrastructure*, a term that not only means they are critical to a nation's economic, financial, and social stability but also makes them increasingly subject to regulation.[110] One cable operator notes that "to a certain degree, all telecommunications infrastructure has been subject to aggression since day one"—the industry has always had to deal with people digging up cables, but today speculation about potential terrorism abounds and rumors circulate about the hiring of pirates to cut cables.[111] In many places, especially in the United States, security measures have been increased to control cable operations more tightly. A group of actors and agencies, collectively known as "Team Telecom," review applications for new telecommunications developments and clear cable projects landing in the United States, and this group has made it increasingly difficult to land a cable on American soil.[112] A manager for a recent cable project told me that it took his company six hundred days to get one particular approval.[113] A cable builder reports that to land a cable segment in the United States, he would have to set up a network operations center in the country as well as a point at which "one person can push a big red button to turn the system off."[114] These requirements make many builders nervous, but they do not have many alternative routes to access the Internet. As one cable builder quips, when laying cable across the Pacific, the United States is still the place to go, since no one has yet figured out the "challenge of how to deal with the Mexicans."[115]

Together these forms of friction—whether inadvertent disruption, natural disaster, or fear of attacks—have drawn attention to the lack of robustness in the cable system, even in spite of the ring networks built during the boom, and have created an impetus in both the public and private sectors to develop more diverse routes. Telecommunications companies have put pressure on governments to address potential sources of friction, stimulating a set of conversations among companies, governments, and regulators (such as ROGUCCI, described at the outset of this chapter) about potential policy developments. This has also motivated a renewed emphasis on security in the geography of route design. The Asia-America Gateway, which began operations in 2009, was developed in response to the Taiwan earthquake: its promoters' main pitch was that it would increase the reliability of traffic between Southeast Asia and the United States by avoiding Taiwan and Japan. Projects such as Arctic Fibre, the SPIN system, and the BRICS cable (which establishes a southern route between South America, South Africa, India, and China while avoiding Europe and the United States) have proposed innovative routes that diverge from the historical contours of cable networks. These developments exemplify Stephen Graham and Nigel Thrift's observation that "disconnection produces learning, adaptation and improvisation" and thus disruptions are often constitutive of new approaches to infrastructural development.[116] However, as the ROGUCCI report observes, cable companies are for the most part content to work "within the current paradigm of incremental improvement based on lessons learned from historical outages": widespread catastrophes have to occur before action is taken to protect the network.[117] Without these catastrophes, potential funders are hesitant to finance improvement projects.

A recent attempt by Pacific Fibre to lay a new transpacific system illustrates the difficulties networks have in developing new geographies, as well as the continuing importance of the state and the rhetoric of security in justifying alternate routes. Since 2001 Southern Cross has been the only cable system linking New Zealand to the rest of the world, a situation that parallels the network geography of the early telegraph era: one company monopolizes the country's international communications infrastructure, meaning that the nation depends on a single system.[118] In 2010 Pacific Fibre announced plans to extend a transpacific system serving Australia that would use New Zealand as an intermediary node.[119] Pacific Fibre was a private company, but it was nevertheless touted as something for the nation to be proud of, the "brainchild of some of New Zealand's brightest entrepreneurs and technology businessmen," and a heroic monopoly buster that would create competition and lower prices for everyone.[120]

From the outset, however, the builders knew that they faced not simply an engineering challenge, but also one of raising funding—a feat other systems had attempted without success.[121] Although Southern Cross was commercially motivated—in part by a lack of capacity—with the upgrades made possible by new technologies, it still had more than enough capacity for the country. Pacific Fibre faced an economic rather than a technological bottleneck. Given that it would be attempting to connect users who already had connections, Pacific Fibre and its supporters used arguments about the need for diversity and information security. The head of InternetNZ, the organization in charge of the country's Internet governance, argued that "the most important thing to us is that another cable would provide a divergent pathway."[122] If Southern Cross were to be disrupted, InternetNZ claimed that it would jeopardize the nation's functioning, especially as most Internet content originates offshore and users' access to basic services is dependent on international links. One telecommunications expert ruminated that if there were a severe disruption at the cable stations, the country's Internet connections could be down for months and "as a twenty-first-century economy, you just can't afford to take that risk."[123] The risk is twofold: not only would there be the economic cost of lost traffic and cable repair, but the country would also suffer a blow to its reputation and could lose its ability to attract technology jobs. "It's a hard thing to talk about," the expert told me, "because you don't want to alarm people about it. And you don't want to alert people to the danger that exists." Mobilizing these fears nonetheless remains integral to any project to develop diverse routes.

Although the ostensible intent of Pacific Fibre was to diversify the cable routes, the historical geography of cable laying presented the company with a number of difficulties. It would be much cheaper and easier to extend a new cable to old landing points. Landing inside existing cable protection zones would make the company's network more secure from other forms of traffic in the ocean, yet this would give the cable network less diversity. It would be difficult to set up new protection zones—this could involve costly and time-consuming fights with fishermen. One cable entrepreneur told me how he responded when the Australian government expressed interest in routing the cable to Melbourne instead of Sydney for diversity (adding extra length and millions of dollars to the network's expense): "The only way it's going to work is if you pay for it. Send check, I'll build. Haven't heard too much since then."[124] The company could not fully diversify the route unless another commercial interest or a government was willing to pay for it, and without an actual disruption to look back to, companies hesitated to put up the money. "When did we last have a terrorist attack here?" the entrepreneur asked me rhetorically. "Never. Who the hell's interested in New Zealand? How could you justify anything here?"[125] Anthony Briscoe of Telecom New Zealand suggests that the resistance to fund infrastructure is a global issue: "As soon as you say 'Internet' now, people think, 'I get it, free.' Yet someone has to make that infrastructure investment."[126]

In addition, to differentiate itself from Southern Cross and to build a case that it would make money, Pacific Fibre sought to plan a low-latency cable—one that would offer a more direct route and less temporal delay. Having the shortest cable route has become more important with the development of the Internet. Microseconds make a difference to groups of users—for example, in banking or gaming—who benefit from real-time interactions. High-frequency traders on global stock markets use computational algorithms to take advantage of the slight changes in price in different locations and to secure trades at slightly quicker rates; companies that have a more direct connection can exploit this for profit. It has been estimated that just a few milliseconds can result in a difference of $20 million in a single month.[127] Once a high-frequency trading company buys capacity on the line, the other traders must do the same to keep up. Even though such low-latency users tend not to purchase a lot of capacity and by themselves cannot sustain an entire network, they are an integral market for new cable projects.

If Pacific Fibre landed its cable in Los Angeles, it could pitch itself as the quickest route to the United States (ironically, Southern Cross mobilized this same argument to get customers to switch to their cable from the Australia-Japan Cable). John Humphrey, part of the original project team, recalls the moment he realized this potential advantage. One night on Google Earth, he drew a straight line between Australia and Los Angeles and was surprised by how close the line was to New Zealand. He suddenly saw that by building from Australia to New Zealand to the United States, his team would have a cable that was quite a bit shorter than Southern Cross, which took more time going up through Hawai'i.[128] The cable's proponents suggested that a lower-latency cable could stimulate a range of new industries, such as making New Zealand a site for data centers.[129] However, this pitch for a low-latency cable—an "express route"—would entail further consolidating the routes at existing locations in the southern hemisphere, landing at Sydney and Auckland as well as Los Angeles. The desire to connect directly to these cities conflicts with the desire to construct a physically diverse network.

Another approach to differentiating the network is to make it simple. Although Southern Cross has many landing points and a complex pricing structure, Pacific Fibre could include only a few points, minimizing potential disruptions and making it easier for customers to purchase capacity. However, the logical conclusion of this argument—made to help finance the network—would be to continue connecting places that were already connected, missing the opportunity to link to locations that are off the grid. The network's potential routes were very close to small Pacific Islands with little infrastructure, such as Samoa, Tonga, Wallis and Futuna, Kiribati, Tuvalu, and Niue. Humphrey described the situation this way:

> We're sitting in the United Nations. I said, "Oh, by the way, we could also land in Kiribati," and they went, "How?" I said, "Well, actually I'm going to have to steer the cable around it. We could pick up those side things, but you see then it gets to be a question of economics because I have investors looking for a return, and if I complicate the system too much or I start putting them more at risk, potential risk—don't know whether it's real or not. Every time you introduce something else that can go wrong you're raising the risk. They may never fail, but one could. Perception is reality."[130]

The landing might not actually include extra risk, but making the network more complicated increases the perceived risk and negates the initial pitch of simplicity—historically established routes are by default perceived as more secure.[131] Moreover, the additional links would make the system cost more: governments in the Pacific Islands would have to pay millions for a spur to be put into their country and to develop local infrastructure with which the cable could connect. If they cannot secure the financing and permitting quickly, within the "aggressive timetable" of the cable project, the cable will not stop there.[132]

Creating an innovative network structure thus introduces a new geography, a set of new things that could fail, which—in the competitive market that has emerged since the 1990s—can deter investment and experimentation. This is especially true in countries that are perceived as politically unstable. One prospective cable builder speculated that during a moment of political conflict, developing nations might hold the cable hostage. He told me about one conversation he had with government officials: "They got a bit offended when I told them this. I just said, 'Your army could take it over,' and they said, 'Well, we don't have an army.' 'I see. Okay, your police.' 'Police?' I said, 'Okay, who in this country has got guns? Someone's got guns. Someone with guns could take it over.' They said, 'Yeah, you're right. Someone could.'"[133] Although this is an unlikely prospect, in the speculative discourse of cable building, perception is reality, as Humphrey noted. These arguments make a difference to private cable builders, who must convince potential customers to sign on and banks to finance the network.

As is the case with many other infrastructures, private cable companies that seek to build an undersea network must build an imagined world around the cable project, using narratives of connection to generate confidence in the proposed network. Charles Jarvie, head of technology and operations for Telecom New Zealand, puts it this way: "It is kind of fake it 'til you can make it sort of an exercise," since anyone can tweet, blog, and spread news about a planned project via social media.[134] Another cable builder recalls: "Your investors are looking at you, going, 'Okay, maybe it is worth it,' and your customers are going, 'Maybe these guys aren't just three guys and a PowerPoint. Maybe they've actually got something.'"[135] Companies have to move past the "perception hurdle," scale "the credibility curve," and transform their work from a mere story "to go real."[136] In this environment, new routes bring uncertainties and expenses, even when diversity and security is a selling point. As the ROGUCCI report notes, "Investors tend to prefer the relative *low risk* option of placing cables along certain existing routes (whose safety is known)."[137] Even if security is a potential selling point for an individual cable system, few companies will invest in a new route to increase the security of the overall undersea network.

In the case of Pacific Fibre, raising hundreds of millions of dollars was a huge challenge, and eventually, the company announced that it had failed to garner enough funding. As one telecommunications head describes the efforts: "There's a story. You're tugging the emotional heartstrings and it sounds good, but ultimately it doesn't get to be funded unless you can convince people that they can stand behind it."[138] Another New Zealand telecommunications worker summarized his perception of what had happened: "[The cable layers] had this little belief that, 'Oh, if every man, woman and child in New Zealand would pay fifty dollars each, they could have as much broadband as they want!' Every man, woman and child in New Zealand is saying, 'Well, I don't want to pay fifty bucks.' Because they think, 'You should actually put it up for free. The government should provide it or someone else should provide it.'"[139] Despite the fact that most people I encountered in the industry agreed that New Zealand could use a second cable (even Pacific Fibre's competitors expressed hope that it would succeed), the arguments for diversity, lower latency, and simplicity—which at many points contradicted one another—were not sufficient to convince investors, governments, or the public to pay for the route.

Pacific Fibre, like Arctic Fibre, sought to avoid geopolitical chokepoints and add physical diversity, creating a more robust and reliable global network as envisioned by the ROGUCCI report. A month after Pacific Fibre announced that it would no longer pursue the project, a new competitor, Hawaiki Cable, revealed its blueprint for a cable between Hawai'i, New Zealand, and Australia.[140] Even imagined cables and failed networks leave a residual impact in the contours of the economic and political landscape—from conversations with lenders to environmental impact studies—that can be picked up later. In addition, the actors who have a say in routing, if not the routes themselves, continue to diversify. Some new networks have shareholders that are also large users of capacity and therefore have a greater say in cable development. Internet companies such as Google have also begun to venture into cable activity.

On the whole, however, there remains a fundamental conservatism in the cable industry that dates back to the telegraph era, and many of the companies remain aligned with, though not beholden to, national governments. One can see this quite clearly in the emergence of Huawei Marine, a Chinese cable supply company that was launched in 2008 to compete with American and French cable suppliers. After Huawei Marine signed a contract to build Hibernia Atlantic's Global Financial Network Project Express from New York to London, along one of the most heavily trafficked routes in the world, the U.S. House Intelligence Committee released a report warning of the risks of using a Chinese supplier and suggesting that their equipment could be used to tap content.[141] Hibernia Networks subsequently halted its work on the cable system and shifted to a U.S. vendor, TE SubCom.[142] The historical emphasis on the security of individual systems and the dependence on established technologies, routes, and players has made undersea cable one of the most reliable communications technologies in the world, yet at the same time it has kept the cable network's geography one of the most static in the history of communications.[143]

Fixing the Cablescape

Chronicling the development of undersea cable paths across the Pacific, from the early transpacific telegraph lines to the latest fiber-optic systems, this chapter has demonstrated how colonial investment, Cold War circulations, and trends in deregulation have shaped the contours of transoceanic cable routes. It has shown that a focus on network topography—the ways that systems become embedded in their surrounding environments—illuminates the dynamics of such historical developments and reveals how Internet infrastructure, as a material set of technologies and practices, has been affected by territorial struggles. In spite of the technological biases that govern existing conceptions of network infrastructure, cable deployment has always been a social as well as a technological process, one that involves negotiating the insulation of one's network and its connection with other systems. The space where our data flows is neither weightless nor easily accessible; instead, it costs large sums of money and often connects places that are already connected (at least under the current economic model). Rather than simply taking a terrestrial route or a straight path between major urban hubs, cables tend to follow routes that are already established.

Throughout history, security has been especially important, yet what is seen as secure has varied over time. In the colonial period, companies used the ocean as a kind of insulation to protect cables from shipping, fishing, and national interventions. Securing traffic entailed extending it through territory that one controlled; the network was diversified by routing multiple cables to the same endpoint. Urban hubs have been variously seen as a source of danger (because they bring potential disruptions) and, in the period immediately after World War I, as a source of protection. In the 1950s cables were made secure by separating systems from urban centers. Because cables often follow existing contours, many of these investments in security remain residual in contemporary network routes.

As the case of Pacific Fibre shows, in the tried and true environment of undersea cabling, it takes a major investment to break free from existing route lines. Although the initial private investment of the 1990s diversified routes, ultimately when private companies—each responsible for its own infrastructure and shareholders—are in charge of network building, there is no one to take a cohesive, systemwide view. As the ROGUCCI report acknowledges, "Investment to ensure the resilience at *a global level* is outside the scope of those building this increasingly critical international infrastructure," and without a reliable cable network, "public welfare is endangered, economic stability is at risk, other critical sectors are exposed, and nation-state security is threatened."[144] This is especially pressing as our networks, which once used wireless transmission as a backup, now need cables to back up other cables. New cables will continue to follow the old routes until governments or other organizations are willing to spend the money it takes to develop new pathways of exchange: global network diversity is a problem that cannot be fixed by market forces alone. Indeed, we can find a historical analogue in the first transpacific cable network, which was also motivated by a need to diversify the network but was not assumed to be profitable by itself. As a result, it was driven by state interest and support. Looking back to this history, we might ask: Who is invested in having a robust global network today? Who will pay for it? What rationales will be compelling enough to convince companies to buy in? At what cultural and historical moments will these rationales become important? As much of our content moves to cloud-based systems, which make us more dependent on a cable network that is always functioning, can we continue to assume that our networks will never fail? And if we accept that we do not have a robust global system, might we instead develop responsible media practices for a precarious infrastructure?

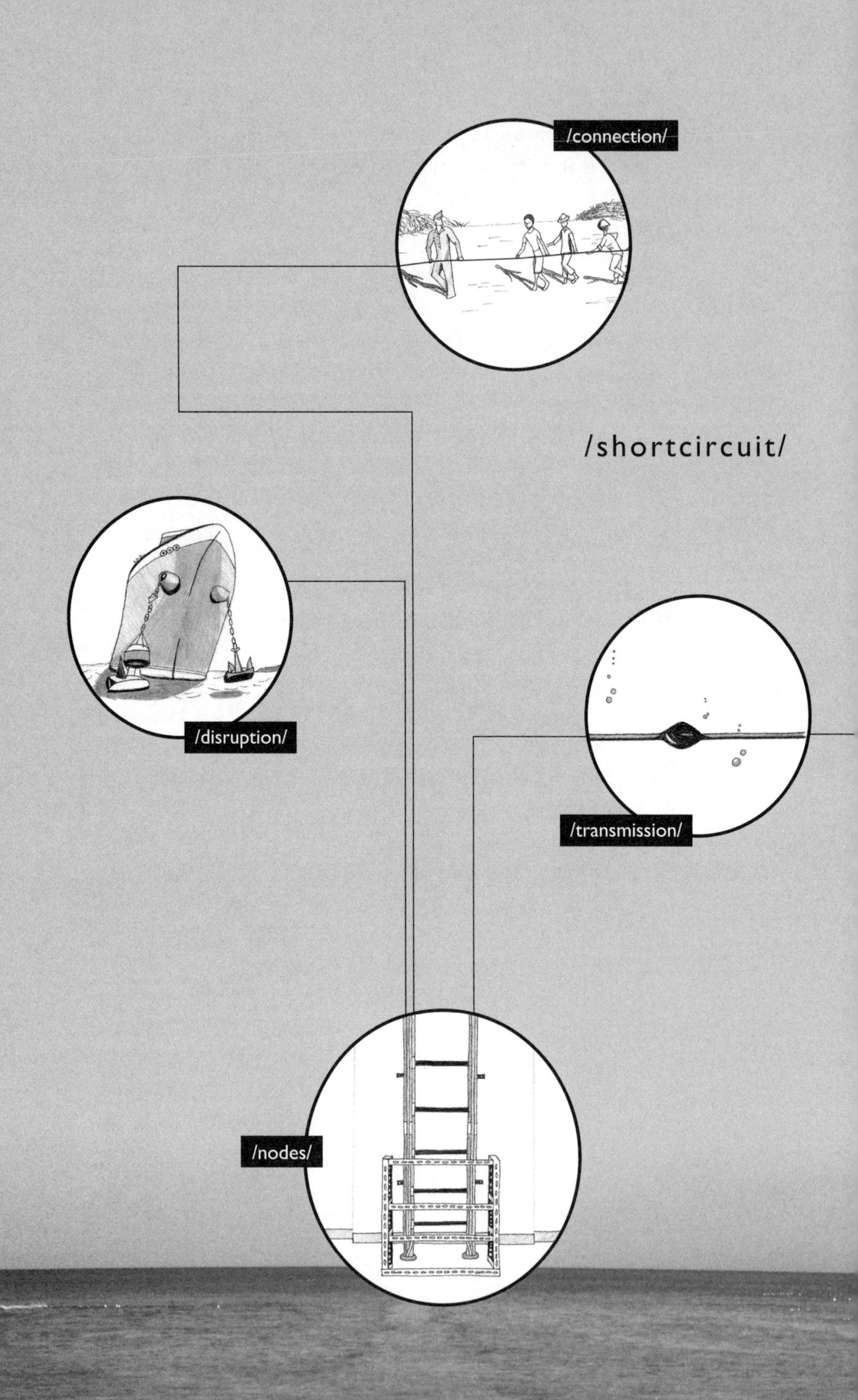
/connection/
/shortcircuit/
/disruption/
/transmission/
/nodes/

TWO

short-circuiting discursive infrastructure

From Connection to Transmission

"Happy is the Cable-laying that has no history."
—Sir William Howard Russell, *The Atlantic Telegraph*

In a series of reminiscences about the first transatlantic telephone cable, the AT&T technician Jeremiah F. Hayes recalls an incident in 1959 when the cable was broken by a Soviet fishing trawler. He writes that the week "started quietly enough. On page 33 of the *New York Times*, there was a brief paragraph announcing that the 'Atlantic Cable is Silenced.' . . . On Monday, February 23 it was reported that the radio circuits were hard pressed during the repair of the cable. Again, the item appeared on a back page."[1] Then, Hayes reports, "things changed," and on February 27 the event was suddenly moved to the front page with an article titled "U.S. Navy Boards a Soviet Trawler in North Atlantic: Action off Newfoundland Follows Breaks in Five Transoceanic Cables" (figure 2.1). The article suggested that the cable might have been severed as a deliberate act of aggression by Soviet ships. The article also used this disruption as

Soviet and U. S. Vessels Involved in Search at Sea

The Soviet trawler Novorossisk, in a photograph released by the Department of Defense

U. S. NAVY BOARDS A SOVIET TRAWLER IN NORTH ATLANTIC

Action Off Newfoundland Follows Breaks in Five Transoceanic Cables

PAPERS FOUND IN ORDER

Admiral Reports Russians Appear to Be Engaged Only in Fishing

Navy statements and relevant documents on Page 3.

FIGURE 2.1. The *New York Times* reports a cable disruption in 1959.

a moment of public education: it contained a wealth of information about the history of undersea cables and how they could be severed. The information was skewed, however, playing up the conflicts of the Cold War and eliding facts that were widely known in the cable industry. Hayes observes that "trawler damage was reported to be unusual. In the context of the Cold War, this misinformation had disturbing implications. The theory was that the Soviet Union would open an attack by cutting cables. The truth was that damage by fishermen had been the leading hazard from the earliest days of submarine cable." In the shift in narration that Hayes calls attention to—from cable break as fact to cable break as narrative event—disconnection becomes visible only once it can be seen as a disruption.

Although they are submerged at the bottom of the sea, undersea cables periodically surface in popular media, with reports of cable landings in newspapers and magazines; visual representations in photography, film, and television; and narratives in fiction, popular nonfiction, and marketing material. These texts shape the public's knowledge about and potential engagement with transoceanic networks. Despite the spectrum of effects that these technologies have had—supporting both democratic interchange and empire building—cable systems rarely surface unless they fit into one of two narrative structures. The first, connection narratives, focus on the design and technological development of an undersea cable: the plot typically unfolds in chronological order, beginning at the point of the cable's conceptualization and ending with its implementation. The second, disruption narratives, describe an unexpected disconnection of the cable and detail the threats not only to transmission but also to a broader cultural order. The range of historical instances considered to be events in plots of connection and disruption is limited. Narratives of both types end at the moment when cable traffic is initiated (or reinitiated), and thus they do not actually depict the cable while it is in operation. Since they fail to attribute significance to operational systems, these narratives actively obscure undersea cables in the public imagination: our inability to perceive cables is structured into the very stories intended to communicate information about them.

This chapter analyzes a range of connection narratives, extending from publicity materials in the 1950s and 1960s to a series of texts produced about the Atlantic telegraph cable during the fiber-optic boom. In these narratives, the technological connection of the cable's two ends becomes the climax of the story: the splice is aligned with a binary change of state from international separation to connection. I argue that these texts, the most widely circulated form of cable discourse, tell us more about the aspirations to certain forms of transnationality, including citizenship in the Commonwealth in the coaxial period and a privatized global sphere during the fiber-optic period, than they do about the operations of cable infrastructure. They are speculative fictions, narrating what the cable might do rather than what it actually does. Narratives of connection can be easily synchronized with both publicity efforts and security: they initially cultivate support for proposed networks and subsequently help us forget that those networks exist. For this reason, the Cold War was simultaneously a time when the modern connection narrative emerged in cable industry films and the operations of undersea cables became more hidden behind the closed doors of the cable station.

In contrast, disruption narratives code disconnection as a broader disruption of globality. Like narratives of connection, they reflect the dominant cultural fears of the time: early stories of the telegraph era traced conflicts to nature, which threatened technological progress; during the Cold War disruptions were coded as external aggression produced by hostile actors; and contemporary discourses often narrate cable breaks in terms of terrorist activity. Even though discourses about cable connection and disconnection have the potential to broaden our knowledge about undersea networks, I argue that the vast majority of these texts do not address active infrastructure, nor do they register the importance of place, local actors, and contingent interactions in shaping these systems. Inscribing cables with cultural ideas about international connection and threats to global democracy, these narratives instead function as a strategy of cable insulation: they are discursive surfaces that transmit dominant ideologies and decrease people's awareness of infrastructure, thus appearing to reduce potential disruption and protect the flows of power routed along cables. By depicting cables as nonfunctional entities without a history of transmission, the narratives limit our possible engagement with these systems. We cannot rely on narratives of connection and disruption alone to convey the significance of cables to governments, companies, or publics that have a stake in their development and operation.

Creating new cultural narratives for undersea cables is critical to an informed public participation with the transnational Internet, especially in a privatized cable system where, as described in chapter 1, public perception can affect the development of new networks. I outline two alternate forms—nodal narratives and transmission narratives—that extend beyond moments of establishment and disruption to portray cables as material infrastructures that must be operated and secured to channel flows of global information. Nodal discourses tell us about a specific site or set of sites in an infrastructure system, focusing on what Lisa Parks describes as the "fields of negotiation that are produced as an effect of infrastructure development and placement."[2] For example, the photographer Taryn Simon's image of the location where undersea cables enter a cable station re-frames the cable landing as a continually occurring process rather than a one-time occurrence. Fictional representations have rewritten the cable in relation to local and regional histories: Gary Kilworth's short story "White Noise" describes the cable station as a place of interconnection among humans, technologies, religion, and the natural environment, and Patrick Downey's novel *Pacific Wiretap* recasts Guam as a hub of transpacific activity, using the setting of the cable station to enhance readers' technological literacy.

Nodal narratives use specific locations in the network to track the intersection of different flows; transmission narratives follow a signal or person across an infrastructure system, tracking movement between interlinked nodes. The trope of transmission is often embedded into longer texts. In several films, from a *Felix the Cat* short (1923) to Krzysztof Kieślowski's *Three Colors: Red* (1994), it has been mobilized to convey an embodied sense of signal movement. In other works, such as Neil Stephenson's "Mother Earth, Mother Board" in *Wired* and Andrew Blum's recent popular nonfiction book *Tubes*, tracking transmission motivates the infrastructure tourist to develop new ways of seeing (and inhabiting) the network. Like nodal narratives, stories about transmission extend the spatiality and temporality of cable discourse to include locations such as the cable station and activities such as cable maintenance. By placing networks in their surrounding ecologies, these narratives can reveal how cables have developed as precarious, territorial, and aquatic entities.

This chapter argues that nodal and transmission narratives short-circuit the dominant discourses of connection and disruption. In technical terms, a short circuit or a shunt fault is a type of cable disconnection in which the insulation is severed, but the central conductor remains partially intact. This process enables power to flow out into its proximate environment, and in some cases allows operations to continue.[3] The point at which the fault occurs is termed the "new virtual earth," a point of interconnection between a global circuit and the local environment, a place where signals are newly grounded.[4] Short-circuiting is an intervention that does not completely sever a connection but instead breaks its layers of protection, feeding power to those around it and leaving the system weakened though still intact. Nodal and transmission narratives puncture the discursive insulation of connection and disruption narratives. They can show cable-using publics the realities of the system's operation, the dynamics of its capacity and use, and the intricacy of infrastructure protection. And in doing so, they help ground the political engagement of cable systems, including arguments for the development of alternative uses and for the production of more resilient and accessible networks.

Connection

Eighty Channels under the Sea (Michael Orrom, United Kingdom, 1962), a short film about the Commonwealth Pacific Cable (COMPAC), a telephone cable laid between Australia and New Zealand in 1963, opens with the following voice-over: "The Commonwealth is a family, a group of people in harmony with each other, who live to the same pattern and assist each other to greater security and good living." We are introduced to users of Commonwealth communications technology: businessmen conducting transactions via telex, journalists sending images via photo-telegram, and families communicating via telephone—all apparently without effort. As if a flashback to their instantaneous communications, the remainder of the film then chronicles the development of COMPAC. The director of the cable's management committee stiffly describes the system's financing. Graphics display the cable's technical design. Viewers are shown the sites where raw materials come from and the factory where the cable is manufactured, in what the narrator characterizes as a "hospital atmosphere." A scene unfolds at Bondi Beach, Australia, when one end of the cable is brought ashore. The system disappears into the water from a cable ship, and the film pauses at moments of tension, such as when a signal amplifier is dropped in. On a New Zealand beach, the cable's two ends are spliced to create a seamless core. As engineers test the circuit, the narrator tells us: "Stage 1 of COMPAC is complete . . . now when the switch girl at the city exchange throws in a plug, distance ceases to exist."

Eighty Channels under the Sea is a narrative of connection: the story's first event is the conceptualization of the cable. COMPAC's completion, motivated by a presumed gap or disconnection between Sydney and Auckland, is the goal that drives the actions of cable planning, manufacturing, and laying. The project is framed as a technological struggle to overcome oceans. When the cable is landed on shore, the narrative climaxes: a technical splice brings international and intercultural connection—in this case, the unity of Commonwealth nations. The end of *Eighty Channels* returns us to the present, to a world where the cable itself, its maintenance and operations, and the distance it occupies cease to exist, or at least cease to have narrative significance. This discursive move—which uses the moment of initiation to shift our attention away from material infrastructure to seemingly immaterial communication practices—make connection narratives indispensable for infrastructure publicity, although they surface in a range of other genres, from children's books to History Channel documentaries.[5] Indeed, *Eighty Channels* was part of a large-scale publicity scheme to generate traffic on COMPAC and to overcome a perceived psychological barrier to making overseas phone calls, given the poor quality of radio transmission.

COMPAC's management committee, whose members came from Canada, Australia, New Zealand, and the United Kingdom, organized the cable laying as an event that would lend itself to the production of connection narratives.[6] To encourage the news media—print, radio, and television—to cover the cable, they announced the dates of cable landings in advance (hoping that local people would attend), contacted journalists and camera crews to document the landings, arranged speeches about the system, and encouraged schools to "adopt" and follow the journeys of Cable & Wireless ships.[7] They launched their own multimedia blitz, publishing photos of the cable stations, distributing booklets with material on the cable's design and installation, tracking the cable ships, and issuing special stamps to commemorate the cable. In this process, the committee members translated a technological feat into narrative and visual terms and reflected on the ways they might stage potentially dramatic moments of cable laying for a mass audience. They argued that "jointing, and laying will provide additional opportunities to create the build-up," and that the landing would be the ideal place for a crescendo of publicity; the climax would be the splice.[8] To celebrate the laying, they staged elaborate opening ceremonies with souvenir programs. Some of the ceremonies were even covered by live television. After the cable began to operate, the staging of events and production of narratives abruptly stopped, leaving only the imagined interconnection of nations.

Connection films formed a cornerstone of the company's publicity drive. Cable & Wireless hired the filmmaker Michael Orrom to write the script for and direct *Call the World* (United Kingdom, 1962), a documentary about the CANTAT cable between Canada and England; *Eighty Channels under the Sea*; and *Ring around the Earth* (United Kingdom, 1964), a thirty-minute documentary on the entire COMPAC system.[9] The films were shown widely in Cable & Wireless's stations and public theaters, as well as at private receptions and film festivals around the world.[10] Their design and narration, like those of the company's other media material, were influenced heavily by the cable publicity committee's decisions.[11] Earlier Cable & Wireless films had focused on technical details, but the committee urged filmmakers to obtain high entertainment values that would leave a "vivid impression" on theater and television audiences, and it cautioned that engineering facts "should be subsidiary to the immediate build-up of traffic."[12]

Such connection narratives, produced in various media formats, were at the time viewed primarily as an advertising opportunity "slanted towards . . . possible traffic raising potentiality," yet the COMPAC committee explicitly recognized a secondary purpose: to articulate the unity of the Commonwealth.[13] The committee warned that in supervising publicity efforts, care should be taken "to safeguard the interests of the [Commonwealth] partnership and to exploit it wherever possible," especially in the filmic texts.[14] The committee policed scripts to ensure appropriate cultural representation. It suggested that "in addition to the Japanese character, there should be a Chinese one to widen the appeal" of one scene and that in another, a reference to Moscow should be taken out so as not to "offend sensitivities" of potential American viewers.[15] It required that scenes be devoted to the contributions of each partner country, including the use of Canadian copper and Australian polythene. In accordance with the committee's requirements, *Call the World* closes with the following words: "one hundred and sixty voices, carried together, but separately, clearly and without interference or fading, on the one cable. Carried across 2,000 miles of ocean. Bringing people together . . . in the future the whole Commonwealth completely linked through the spoken word." As in *Eighty Channels under the Sea*, once the cable is connected, there is no further discussion of the system's materiality or geography—the technical splice shifts the narration to intercultural links and the next planned cable.

Throughout the 1950s and 1960s, the telecommunications industry coordinated the production of cable narratives across media and national boundaries—similar tropes surface in a range of different cable promotional materials.[16] *Cable to the Continent* (United States, 1959), produced by AT&T, frames the transatlantic telephone cable from Canada to France as a project of international cooperation between the United States and Europe. The film cuts between cable workers from each partner country and then to telecommunications company buildings in France, West Germany, England, and the United States. At the film's conclusion, over images of operators working the cable (a brief moment signifying operational infrastructure), the narrator concludes that the cable will "bring many nations closer together, both politically and economically." Even though its transmissions "will be from many nations in many tongues," he tells us, "the cable itself speaks a single language to all, that is the language of friendship and cooperation between the men and women of France, West Germany, England, and the United States, who conceived and brought to completion the cable to the continent, man's newest memorable victory over distance and the sea." As in the Commonwealth films, the moment the cable starts to function, the discourse shifts to aspirations of international cooperation—an ever-present trope during the period after World War II—and the operational cable remains a structured absence. Even though cable geographies become briefly visible, the actual functions of the system, operations that were increasingly hidden from view and sheltered underground, are excised from the story. We do not see what goes on behind the cable station's closed doors, the signals that actually crisscrossed the wires, or the measures taken to maintain and protect cable systems. These processes remain sheltered, insulated from our view.

Connection narratives produced shortly after the transition to fiber-optic cables substitute for a message about international cooperation the imagined form of transnationality in the 1990s: free-enterprise capitalism dominated by American investment. Between 1998 and 2006 a cycle of texts was produced about the original Atlantic telegraph cable, ranging from popular histories such as John Gordon's *A Thread across the Ocean: The Heroic Story of the Transatlantic Cable*; to television documentaries including *Transatlantic Cable: 2500 Miles of Copper* (Yann Debonne, United States, 2000) and *The Great Transatlantic Cable* (Peter Jones, United States, 2005); and novels such as John

Griesemer's *Signal & Noise*.[17] Many followed the structure of the connection narrative, beginning with the conceptualization of the cable and extending to its initiation. Although these texts describe the laying of an early telegraph cable, the technology is framed as significant because it forms a historical precedent to contemporary fiber-optic networks.[18] *A Thread across the Ocean*, for example, ends with the statement that, in establishing the Atlantic telegraph, Cyrus West Field "laid down the technological foundation for what would become, in little over a century, a global village."[19] Drawing a connection between new and old, these texts implicitly narrate the dramatic extension of undersea fiber-optic networks, which increased during the same period from 1.48 million to 45.1 million voice paths across the Atlantic.[20]

Instead of highlighting the coordination of nations, these narratives tend to situate agency in the heroism of a single American entrepreneur, Field, and they intertwine his personal rise to success with the completion of the cable itself. *A Thread across the Ocean* chronicles Field's fight as he pushes on, despite large odds, to secure financing for a transatlantic cable: his personal enthusiasm and drive is the narrative's primary causal thrust.[21] When not tracking Field's contribution, the book focuses almost exclusively on other heroic individuals, including the engineers William Thompson and Charles Bright. The narrative's structure is oriented around each of these characters' rise to greatness, and dramatic moments are generated by their breakthroughs: singular heroes rather than coordinating nations produce global communications technology. In many of the Atlantic Cable narratives, Field's work as an entrepreneur is the key contribution. In *The Victorian Internet*, Tom Standage writes: "Nobody who knew anything about telegraphy would be foolish enough to risk building a transatlantic telegraph; besides, it would cost a fortune. So it's hardly surprising that Cyrus W. Field . . . was both ignorant of telegraphy and extremely wealthy."[22] A lack of technological expertise and a talent for fund-raising leads Field to success.

The narrative of Field as entrepreneur is often paired with a celebration of U.S. influence on the project. *The Great Transatlantic Cable*, an *American Experience* documentary produced by PBS, suggests that like the cable deep in the "sludge" of the ocean floor, Field, "a man once called by historians the greatest American," is now unknown by most Americans. The documentary opens with the narration of an early transatlantic attempt: a sequence of images shows the USS *Niagara* sailing, an American flag flying above, and the ship's "extraordinary cargo," a thousand tons of cable coiled below deck. It cuts to a young man standing on the cable, looking up to the ship's hatch. His hands are firmly on his hips, and sunlight shines down on his face. The narrator says: "The vessel also carried a thirty-eight-year-old self-made millionaire named Cyrus Field." The aural and visual transition suggests that Field, as much as the cable, might be considered the ship's precious cargo. Left completely out of this opening scene is the much more dramatic story of the other half of the cable, which—carried on the British ship *Agamemnon*—hit huge storms and became severely tangled. The fact that almost all of the engineering, resources, and expertise came from the British is hardly acknowledged. Following the *Niagara* and Cyrus Field, the film instead frames the cable laying as an American endeavor and a crucial part of U.S. history.

Although Field's capitalist drive is portrayed as the primary causal force, which makes the cable a distinctly American success, the conclusion of these Atlantic Cable narratives is almost always framed as the overcoming of international tensions between Britain and the United States and the subsequent establishment of global and democratic networks. *The Great Transatlantic Cable* closes by telling us that the project "helps to cement Anglo-American culture and it helps to create really what you might call the Victorian world, which is united by this cable." Like COMPAC's connection narratives, the Atlantic Cable narratives almost always end at the moment the cable is activated (leaving out decades of its operational life) and assume a smooth transition to peaceful international interactions, the foundation for today's equitable global village.

As the narratives about the Atlantic Cable emphasized the unity brought about by the singular American entrepreneur, they reflected the heightened individualism and neoliberal practices of the 1990s and legitimated a new set of approaches to laying cable in the fiber-optic era. As described in chapter 1, after the deregulation and privatization of the telecommunications industry, independent companies and entrepreneurs—most of whom were based in the United States—were breaking with historical monopolies and seeking independent financing for new cables. Private companies found in the transatlantic telegraph story a model to articulate their distinction from the national alignments that had long permeated the industry. Hibernia Atlantic, an undersea cable company, sponsored and hosted a black-tie event at the New-York Historical Society in 2008 to commemorate the first transatlantic telegraph cable. The event featured a speech by the great-great-grandson of Cyrus Field and resulted in a number of news articles that linked the company's work on fiber-optic systems to Field's endeavor. The cable construction company TyCom helped fund a Smithsonian exhibition featuring the Atlantic Cable. Histories of the cable continue to appear regularly in today's industry materials. They provide a rich narrative and iconographic library for companies to use in representing the origins of a private cable industry and cultivating support for further capital investment.[23]

By narrating cables as infrastructures developed by heroes and American capital, Atlantic Cable texts avoided a long history of the national control of cable systems, international cooperation, and extensive investments in (British) technology. By suggesting that entrepreneurial efforts led to a democratic, progressivist system that smoothed out conflicts, they also avoided the long history of the use of technologies to reinforce unequal power relationships. One notable example that counters this trend is the novel *Signal & Noise*, which critiques the Atlantic Cable story's reinforcement of capitalist ideology, heroic construction, and democratic global communication. Fictionalizing the transatlantic cable laying, *Signal & Noise* replaces Cyrus Field as a central motivator for the story with an engineer, Chester Ludlow. As the narrative begins, Chester is uninterested in the public dimension of the cable (its story) and merely wants to work on its design, but as the story unfolds, he is coerced to narrate a Phantasmagorium show in order to help raise money for the system. Initially reluctant, Chester is soon transformed into the cable project's celebrity. The newspapers and the cable consortium frame Chester as a hero connecting two nations, but the novel makes clear the deliberate construction of this heroism. Contemplating the cable's opening ceremony, one investor dictates a song about the endeavor to his assistant: "The cable lies under the ocean . . . Thus Ludlow brought England to me!" His assistant stops him and asks, "'*Ludlow* brought England,' sir . . . . Isn't it Mr. Field actually who is chairman of the syndicate?" The investor answers, "'Mr. Field' doesn't scan. . . . Besides, Ludlow's the man they're all going to fall for. . . . He has the look. The fair-haired son. The matinee idol. Write down 'Ludlow.'"[24] The cable show is framed as an act of artifice, a trick to get people to gamble their money. As Chester is caught up in his role as the "connector," he loses the sense of purpose with which he began the expedition. After the two cable-laying attempts in 1858 and 1865, Chester chooses not to go on the ship that carries the successful cable of 1866. The novel's plot leaves out this final journey, and in doing so, leaves out the moment of technological connection.

The collapse of distance that Chester proclaims the transatlantic telegraph will bring does not enable characters to connect on equitable terms but instead pulls them apart. Chester's work on the cable draws him away from his wife, and he has an affair. Even his new lover questions what kinds of signals the cable can really transmit, telling him that "all your work on these cables, and still all anyone can send is signals in a code no one can speak."[25] Throughout the novel, there is an uncertainty that dots and dashes can be significant in themselves, without a code that somehow translates them into meaningful interconnections. *Signal & Noise* depicts other kinds of signals, those emerging from the depths of the ocean and from a spirit world beyond, as equally valuable. If histories of the Atlantic Cable at the turn of the twenty-first century narrate how grandiose aspirations of heroic men, technology, and capitalist ideology overcame the ocean in service of democracy, *Signal & Noise* challenges these assumptions and suggests that what we think is noise might actually be signals occurring in the absence of decoding mechanisms—signals written in a "code no one can speak." Unlike existing narratives of cable connection, the novel questions the cable splice's unproblematic initiation of international connection. It instead asks readers to attribute meaning to various kinds of transmissions sent along the cable, and to account for the range of different worlds that they might interconnect.

Describing an early telegraph voyage, Sir William Howard Russell writes: "Happy is the Cable-laying that has no history."[26] Russell suggests that the cable-laying project is better without drama and conflict, and therefore without history. In contrast, connection narratives reveal to us that although cable *laying* may have a history, cable *operation* should not. As they ignore the active infrastructure and position the operational cable as a phenomenon that smoothly functions independent of human agency, connection narratives are fundamentally limited as a foundation for ongoing discussions—whether by technologists, policy makers, or communications researchers—about undersea cables. These narratives assume that broader and ideologically inflected intercultural and economic connection, whether that of a multicultural commonwealth or the American democracy, follows directly from technological connection. Stripped of all history after the splice, the plot excludes the work that is constantly needed to keep the signal going, efforts that take up much of the cable companies' time and money. In the narrative of connection, there has been little consideration of who uses the cable, how revenue is generated, how traffic is coordinated, and what entities pay for things like equipment upgrades, labor and training, security, and system maintenance—all concerns that continue to be pressing for cable operators today.

Disruption

Narratives of disruption are generated by an actual or potential cable break, an event that shifts the information environment from a state of interconnection to one of disconnection. The core tension is a struggle of cable companies with natural or social threats, and conflict is resolved with the cable's repair or

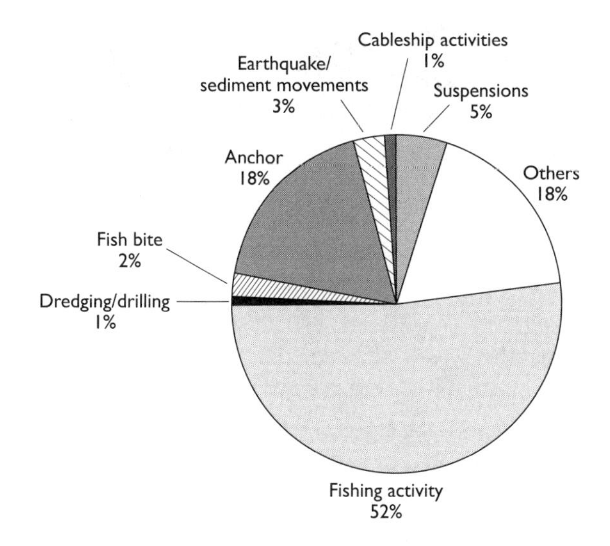

FIGURE 2.2. A recent assessment of the sources of cable breaks. Data from Thomas Worzyk, *Submarine Power Cables: Design, Installation, Repair, Environmental Aspects* (Berlin: Springer-Verlag, 2009).

protection. Plotting the cable beyond its initial activation, often in a series of news articles, disruption narratives have the potential to stimulate discussion about operational networks. They could treat topics such as cables' susceptibility to environmental factors (from tectonic shifts to the interference of fishermen) and the role of local actors in global systems, since human activities can be credited for 70 percent of faults (although systems are only rarely disrupted by a direct attack) (figure 2.2).[27] Disruption narratives might also convey the regularity of cable faults: though accurate reporting is difficult, faults are a routine part of global cable operations, occurring on average once every three days.[28] Although they do not always result in a complete disruption of traffic, faults often mean that information must be routed along alternative cables, a process that can cost cable owners millions of dollars in lost revenue. There is an entire marine maintenance industry that has to be funded by cable owners even in the absence of breaks. Lastly, narratives about disconnection might point to the importance of failure in developing dynamic systems and help us understand the impacts that cable breaks have on the industry, cable development, and users.

Instead of using the cable break as an opportunity to increase technological literacy, popular discourses downplay the diversity and frequency of faults, more often coding disconnection as a large-scale disruption to the dominant cultural order. Since the beginning of cable laying, the most frequently cited threat to cables has been the ocean and, by extension, nature situated in opposition to modern cable technology. In Rudyard Kipling's poem "The Deep-Sea Cables," he describes the telegraph cables' seafloor environment:

The wrecks dissolve above us; their dust drops down from afar;
Down to the dark, to the utter dark, where the blind white sea-snakes are;
There is no sound, no echo of sound, in the deserts of the deep;
Or the great gray level plains of ooze where the shell-burred cables creep.[29]

Kipling emphasizes the distance between the black, silent deep and the white cables carrying people's words, as well as that between the ocean surface where human activity takes place and the "waste" of nature that threatens to dissolve it. This opposition between cables and the depths of the ocean was fundamental to early cable discourse. At a dedication ceremony for the first transatlantic telegraph, U.S. Senator Edward Everett commemorated the cable, describing "the thoughts we think up here on the earth's surface, in the cheerful light of day" that then travel "far down among the uncouth monsters that wallow in the nether seas, along the wreck paved floor, through the oozy dungeons of the rayless deep."[30] Cables, capable of transmitting thoughts and senses and often seen as white or light, were juxtaposed with a senseless and dark environment populated by sea monsters that might seize them.[31]

The threatening element in this environment was often seen as water itself, a material that could intercept the electric signal and break the circuit. For example, in a 1930s newsreel about a telephone cable under San Francisco Bay, a voice-over for images of men working on a cable ship describes the effort "to combat a small but deadly peril to this circuit . . . a single drop of water." The narrator tells us that "it is of vital importance to prevent the entry of a single drop, for if one drop can enter so can another, and another, and another, until the electrical transmission of speech is impossible!"[32] A graphic reveals the layers of protection enwrapping the cable. The narrator concludes: "So look out Father Neptune, down there in your underwater kingdom, here come 1,056 wires, there's no use trying to get them wet, for I don't think it can be done."[33] In this video, there is no elaboration of the many human obstacles to the cable's functioning (the "accidents"), or to the fact that underwater routes were often used to enhance security for cables. Rather, the potential disruption is generalized and mythologized as "Father Neptune." Another story from this period about cable repair describes it as a "truly Dantesque tableau of the contest between Nature in a savage mood struggling to maintain, and Man in his tenacious efforts to subdue her."[34] The trope of ocean as threat continued into the coaxial period. In Arthur Clarke's cable history, *Voice across the Sea*, the laying of undersea cables is understood first and foremost as a struggle against the ocean. Clarke writes: "This is the story of man's newest victory in an age-old conflict—his war against the sea."[35] Instead of narrating the ocean as a site of technological development that is politically valuable to nations and one that in fact insulates cable traffic, these discourses of disruption frame the ocean and water in general as forces of nature that threaten to dissolve human communication.

Through the 1940s and 1950s, there was a broad transition in the cultural conception of underwater spaces. Rather than simply a natural or wild domain overseen by Father Neptune, oceanic space became widely surveilled and militarized, an extraterritorial space where conflicts played out during World War II and the Cold War.[36] In this period cable protection was articulated and formalized as a struggle against possible aggressors. In 1958 the Cable Damage Committee (later the International Cable Protection Committee) was formed to address marine disruptions. The romantic drama that positioned undersea cables against a wild nature slowly gave way to discourses of security, and threats to undersea cables were framed as deliberate "external aggressions" in which sea animals' animosity paralleled that of enemy nations.[37] A short article in *Underseas Cable World*, titled "The Natives Are Unfriendly (Sometimes)," discusses the "attacks" inflicted by marine animals "prowling blindly" on the ocean floor, with vocabulary mirroring the language used to describe submarines.[38] This rhetoric also pervaded the *New York Times* article mentioned at the beginning of this chapter, in which a fairly "routine" transatlantic cable break became visible only when it could be narrated in relation to a potential attack by foreign nations. Rather than nature's disruption of technology, this break was narrated as a Soviet disruption of American space.

In the fiber-optic period, cable breaks continue to be written about in relation to security concerns, yet they are depicted as potential targets for pirates and terrorists rather than nations. When the Vietnamese government permitted fishermen and soldiers to pull up undersea cable laid before 1975 to sell it as scrap, their plans went awry: fishermen pulled up two of the country's three working fiber-optic lines (a significant decrease, given that 80 percent of the country's information traffic was carried by undersea cable).[39] The government rescinded the permission it had given to salvage copper lines, began monitoring the route of the last cable, arrested the cable "thieves," and planned an enlightenment campaign to inform fishermen of cable locations and the dangers of cable theft. On the one hand, news about the event stressed the fishermen's lack of knowledge about the cables and their locations. Some reports suggested that fishermen were not able to tell the difference between copper and fiber-optic cables.[40] The development of an enlightenment campaign suggests that the fishermen may have unintentionally pulled the fiber cables up to begin with. On the other hand, these narratives suggest that the pirates deliberately threatened the global network. Newspapers reported that the Vietnamese government had imposed the death penalty on the perpetrators, citing interference with national security projects as the infringement.[41] In these reports, breaks and repairs are not situated as part of a broader historical conflict dating to the 1800s, but as an example of contemporary threats of theft and piracy.

Another series of cable breaks occurred in January 2008, when two cables in a row were mysteriously snapped near Alexandria, Egypt. This is one of the pressure points for global undersea networks with few alternative traffic routes, and the breaks caused an immediate decrease in Internet connectivity—reportedly up to 75 percent between Europe and the Middle East.[42] Ships' anchors were initially blamed for the damage. Then, on February 1, a third cable was damaged west of Dubai, and rumors quickly began to spread that terrorists were responsible for the disruptions. Seeming to corroborate this view were Egyptian authorities' reports that there was no video footage of shipping in the area at the time of disruption. A host of online blogs and news articles speculated about the disruptions' possible causes. One article began: "Was it a ship's anchor? A lurking spy ship? A shark with peculiar tastes? Just who was behind the recent cutting of several fibre optic cables in the Mediterranean a few weeks ago? The mysterious episode cut off nearly 73 million people from their Internet, telephony and other services—and now forms the basis of the hottest conspiracy theory."[43] The discussions failed to engage cable history and educate the readers as to the variable impact of shark bites (a problem in the 1980s), fishing anchors (a threat throughout history), and terrorist attacks (none known as yet). An article in the *Economist* later that week criticized the discussion as "online frenzy that seems way out of line" and reported that "it may be rare for several cables to go down in a week, but it can happen. Global Marine Systems, a firm that repairs marine cables, says more than 50 cables were cut or damaged in the Atlantic last year."[44] This brief mention in the *Economist* highlights the precariousness of communications cables and provides a small window into routine cable maintenance challenges. Regardless of the actual cause, as in the case of the Soviet trawler, these breaks only surfaced in relation to fears about broader global disruptions—in this instance, the potential disruption by terrorists in the Middle East. In turn, the perception of a potential terrorist threat has helped justify new forms of oversight by governments.[45]

Like the narratives of connection, disruption narratives tend to chronicle a limited set of events, moving from a state of disconnection that is remedied

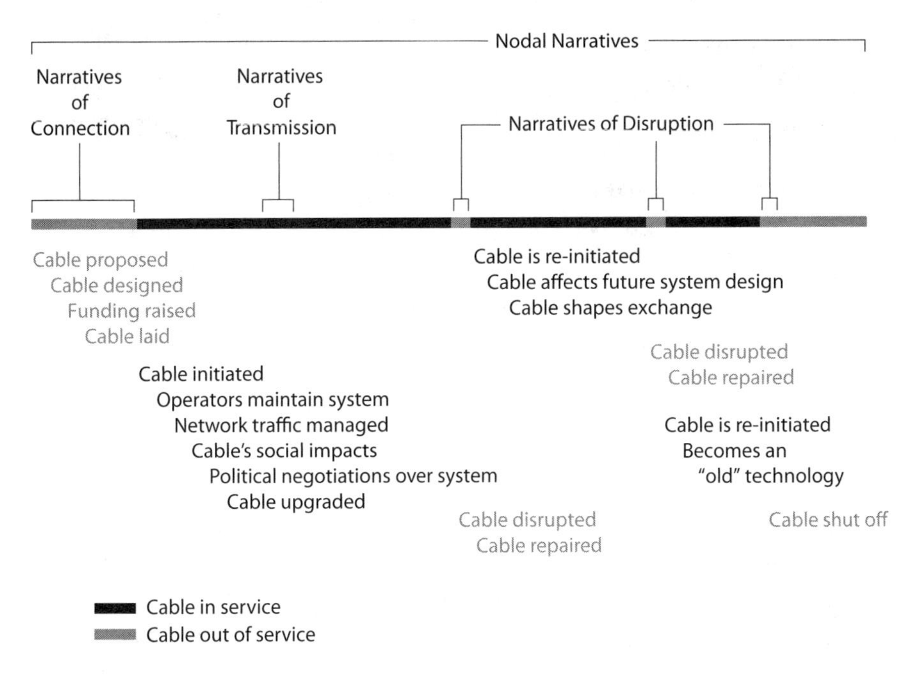

FIGURE 2.3. The temporal limitations of narratives of connection and disruption.

by a cable's initiation and ending with the disappearance of the working cable. After the undersea networks are repaired, the discussion about them subsides. Although connection narratives tend to ascribe agency to specific companies (in the case of COMPAC) or individuals (in the case of Cyrus Field's Atlantic Cable), disruption narratives often displace agency to generalized and perceived threats to global interconnection: in the telegraph era, nature was the primary obstacle; in the 1940s and 50s, the threat became a hostile nation; and in the fiber-optic era, cable disruption is linked to terrorism and piracy. The most recent fears do not completely replace the former ones but rather are layered onto one another, so that in narratives about today's fiber-optic cables, the ocean and its waters are still portrayed as threats. All the while, most disconnections continue to be caused by fishing nets and boat anchors.

This focus on cables at moments of initiation and disruption, and the rare visualization of cables as active and operational infrastructures, reinforces the invisibility of cable systems in the public sphere. In discourses of disruption, local and contingent events and actors such as fishermen are not discussed in detail: they do not make a good story. When protection is enacted and the cable is repaired, the story ends and cultural fears about threats to connection are temporarily relieved. Even though these discourses appear to expand the frames through which we perceive undersea cables and the spaces in which agency can originate, like the narrative of connection they function as discursive strategies of insulation that blind us to the material infrastructure (figure 2.3). They obscure the historical layering of infrastructures, the ongoing policies that regulate their operation, the people who use their capacity, the struggles to protect them through negotiations with local actors, and attempts to build a broader and more diverse network. In the second part of this chapter, I turn to two alternative modes of representation that grapple with the cable as it is in operation and that, in doing so, fill in the gaps in narratives of connection and disruption.

Nodes

Standing at the individual nodes of our networks, we can see the conflicts, contestations, and negotiations that shape systems on the ground. Take, for example, the cable that runs directly over Gun Beach, in Guam (figure 2.4). Tourists step on its conduit without looking twice, intent on getting to the ruins of an antiaircraft gun on the other side of the beach, an installation constructed by the Japanese during World War II. Guam had strategic military value in part due to its position as a cable hub. At the other end of the beach is the Hotel Nikko, one of the early Japanese five-star hotels in the area—a structure dependent on the island's advanced communications and transportation infrastructure. Undersea cables, shuttling signals to the United States, Australia, and Japan, bisect a material reminder of the Japanese wartime attack and one of the largest indications of Japanese development on the island, two sites with which communications traffic have been intertwined. The cable conduits themselves are worn down and blend in with the rocks on the beach. In the water their surfaces look smooth but rippled, like the ocean. In them, we can see traces of the atmospheric and geological currents that have long positioned the island as central to transpacific movement. Gun Beach is a node in the cable network, a point of interconnection and redistribution—not simply of physical signals, but of transoceanic cultural and political currents.

To tell a nodal narrative is to describe a particular location, such as Gun Beach, as it is intersected by different kinds of connective threads over time.[46] Nodal narratives can be evoked in a single image, such as Taryn Simon's photograph "Transatlantic Sub-Marine Cables Reaching Land," which reveals the site where cables enter a transoceanic cable station. Five cables run from the top of the photograph directly down the center of the image to intersect the floor, where they are surrounded by a metal frame. The caption states the location

FIGURE 2.4. A cable conduit extends out to sea at Gun Beach, Guam.

of the cable landing (at VSNL International, Avon, New Jersey), their large capacity (sixty million simultaneous phone calls), the geographic spread of cable systems (all continents except Antarctica), and details about cable laying and protection (referring to shore armor that keeps them safe from sharks and fishing). Part of Simon's photography series titled "An American Index of the Hidden and Unfamiliar," the picture is viewed next to a range of sites obscured from public view, including a nuclear waste site, a U.S. Customs and Border Protection Contraband Room, and a Research Marijuana grow room, all places that Simon states are "foundational to America's daily functioning and mythology" yet remain "out-of-view, overlooked, or inaccessible to the American public."[47] In this approach, the cables are first marked as "hidden," yet the image renders them as bare, seemingly accessible, and surrounded by open space. The gate around them does not appear to be an adequate mode of protection—anyone who approaches might easily reach over it to touch the cables. Since they are discursively marked as off-limits, we must assume that the cables are protected in some other way. The imaging of this open space thus draws our attention to the invisible structures of security that remain outside of Simon's frame. Although it shows us the cables themselves, the photograph also directs our attention to the lack of information we have about signal transmission.

FIGURE 2.5. A traditional image of a cable landing at Vatuwaqa, Fiji, in 1962. Courtesy of Archives New Zealand. COMPAC Commonwealth Trans Pacific Submarine Telephone Cable [Archives Reference: AAMF W3327 457] Archives New Zealand The Department of Internal Affaris Te Tari Taiwhenua.

Focusing on active cables at the landing point, this image counters the dominant understanding of what it means for undersea cables to land. The cable's beach landing has been the single most commemorated event in undersea cable history. Traces of landings can be found in scrapbooks, home movie footage, companies' publicity websites, and historical texts, and cable companies often represent the landing as part of a narrative of connection. As documented above, cable landings have also historically served a cultural function in the celebration of the nation and ideologies of technological progress.[48] To document or be present at a cable landing is to be a witness to the coming together of nations or worlds. Public participation is allowed in part because the cable is routed through public space; at early landings, locals helped pull it ashore. Images of cable landings have been almost identical for the past 150 years (figure 2.5). A large cable ship is typically visible from shore. A long line of buoys, made of inflatable floats or drums, are strung out from the ship and hold up the cable below. These photographs use similar techniques for framing: the cable extends from the line of the horizon to the shoreline in the foreground, both of which run roughly parallel to the frame itself. There is a strong sense of directionality: the cable moves toward land and the viewer. Like nodes in a network, the inflatable floats are the only visible evidence of the cable. When the strings to the cable are cut and the balloons are freed from the cable's weight, they move up, often literally into the air. This is the moment when communication is liberated from material constraints.

"Transatlantic Sub-Marine Cables Reaching Land," like all landing images, features the cables on the *y* axis, intersecting the horizontal line of land that runs parallel to the bottom of the frame. However, the image makes it unclear where the cable reaches land. The history of cable-landing images would encourage any viewer familiar with their aesthetics to see these cables as extending down from the ceiling to reach the ground, or land, below. This is a reading encouraged by the image's title: the cables are "reaching land" rather than coming into the station. In reality, the cables reach land outside of the station and run under the ground, then into the station, up the wall, and across the ceiling to different access points. This is a contradiction: the visual point at which the cables land in the photograph is not the material point where signals actually reach land; rather, it is the point at which they leave land. Simon's photograph reverses the sense of directionality that so strongly orients most landing images. The area from (or to) which the cables run is undisclosed, cropped out of the frame. Although landing pictures commemorate the moment when the cable reaches shore and the resulting dissipation of its materiality, Simon's image visualizes cables' emergence from one unknown space into another. Prompting the viewer to infer what has been left invisible (the building that the cables emerge into, the direction they come from, and who the metal gate is meant to protect the cables from), the photograph directs attention, even if it does not provide answers, to the persistence of the cables' materiality after the moment of connection as well as to the secrecy that surrounds these systems.

Kilworth's "White Noise" performs a similar operation, using the node as a jumping-off point for contemplating the unknowability of a network. And, like *Signal & Noise*, "White Noise" leads readers to question what it is possible to perceive at all. Set in the Middle East, the story follows two telecommunications engineers as they journey to perform routine operations at a haunted cable station on the Red Sea. Arriving at the station, they find a speaker hooked up to the coaxial network's circuit, broadcasting a live feed from the cable's deep-sea repeaters to the surrounding environment. Animals around the station have gone silent as they listen to the feed. The engineers realize that they are hearing noises from the past trapped in cold currents that flow over the bottom of the seafloor. In "White Noise," the cable station is not simply a site of interconnection for circuits. It is also a place where the technological networks of the present connect the engineers, and the reader, with voices from history.

Listening to the past, the narrator thinks he hears a battle and attempts to provide technically plausible explanations: "I *could* hear faint shouts and yells, under the rush and hiss of the white noise. Perhaps, perhaps the rumbling of wheels, the rattle of metal . . . ? I tried rejecting these ideas and replacing them with the thought of trawler nets snagging the cable, or predatory fish worrying the cable."[49] He soon realizes that he is listening to Moses crossing the Red Sea and is about to hear the voice of God. His co-worker abruptly turns off the machine, proclaiming: "We mustn't hear it. . . . If we were to hear God speak, we would *know*. We would know for sure. It would make a worthless thing of faith."[50] "White Noise" depicts the cable station not only as an operational site where periodic maintenance must be conducted, but also as a critical node and a location of transfer between apparently incompatible worlds: the world of religion and the world of technology, the natural environment and the human environment, and the past and the present. As in Simon's image, the reader is not given an understanding of the dynamics that govern such transfers between worlds. The narrator, consumed by his desire to identify the sound's source, struggles for control of the station's connection, but his friend sets fire to the cable station, ruining the equipment. The node, a gateway to the past, is deliberately destroyed in order to keep comprehension about these networks beyond the characters' grasp. In turn, the story imbues the station with a sense of mystery and prompts the reader to ponder the hidden operations of the cable network.

Although "Transatlantic Sub-Marine Cables Reaching Land" and "White Noise" reveal the network's access points only to obscure the network itself, *Pacific Wiretap*, written by the telecommunications expert Patrick Downey, is a nodal narrative that conveys substantive technological information about network operations to heighten the drama of a crime story. The novel focuses on the adventures of a young engineer, Jonathan Fox, who is sent to Guam in order to investigate a wiretapping scheme. As in "White Noise," much of the story takes place inside a transoceanic cable station. However, as Fox moves in and out of the cable station, Downey educates the reader about the interior workings of the undersea network, the jobs of maintenance and operations, and the ways that criminals might use a cable to communicate free of detection. In this book, the skilled technical engineer's desire to understand the network's operations is satisfied rather than thwarted, and Guam's cable station becomes a site for educating readers about transpacific networking.

Just as Simon's photograph points us to the embeddedness of the hidden cable landing in American culture and "White Noise" draws connections between the network and the Red Sea, the criminal cable scheme being set up in *Pacific Wiretap* entangles us in Guam's local history. In the book, an air force captain, two network technicians, a Japanese war veteran, and his grandson plot to steal $75 million in gold that had been plundered from the Philippines during World War II and buried on Guam by the Japanese military. This event had been made possible by a set of earlier connections traversing Guam: its colonization by the Spanish and its ties to the Philippines, the Japanese invasion, and the American military power on the island. As Fox unravels the mystery—all by listening in at the cable station—the reader becomes acquainted with Guam's political history, including its part in the conflicts of World War II, its current military dominance, and its economic proximity to Japan. The cable station is, as in "White Noise," a place at which a connection between past and future can be discerned, where global extensions are locally mediated and at times intercepted.

Although narratives of connection and disruption describe the transcendence of distance by undersea cables, nodal narratives take the persistent materiality of undersea networks as their starting point, bringing the reader into contact with cabled environments. Simon's photograph asks us to see the cable landing as a continual process rather than a historical climax, and as a place where American power is exerted across the network. "White Noise" and *Pacific Wiretap* prompt us to see cable stations as not only conduits to a technological system, but also points where networks become intertwined with historical events and regional conflicts. In nodal discourse, the narrative tension is generated by the intersection of and conflict between these various movements. Together, the narratives extend both the temporality of cable representation, insisting that the cable continues to have a story even after it lands, and its spatiality, introducing us to the core geographies of an active cable system: the route, the station, the landing, and the island.

Transmission

Following a signal has long been a compelling way to represent a network. One manager at Cable & Wireless suggested in the 1940s that it might be a good publicity pitch, since "there is a world of romance contained in the few minutes during which [a signal] has been flashed across the floor of the ocean, through submarine forests, among hidden hills and valleys, over regions where the sun's rays have never penetrated and which human eyes have never seen."[51] Tracking transmissions, like nodal discourse, expands the times and spaces covered in a cable narrative's plot. Rather than focusing on a single node, however, discourses about transmission move between different cabled sites. Take, for example, the opening of Kieślowski's *Three Colors: Red*, which traces the signals of a single telephone call as they move through underground ducts, cable landing points, into and out of the ocean, and fail to get through, producing a busy signal. Or consider the documentary *The Spy Factory* (C. Scott Willis, United States, 2009), which follows a hypothetical e-mail message embedded with National Security Agency keywords such as *biological warfare* from Malaysia, through undersea cables to China, under the Pacific to a California landing point, and to a cable station in San Luis Obispo. Data traffic is digitally projected over these landscapes, giving us a layered image of physical and information environments. As they track our signal traffic, these short narratives of transmission reveal the beaches, forests, valleys, and underground environments through which cables are routed, giving a sense of the network's nodes as well as the operations that sustain cable connections—even if, in the latter case, they are significant only as places in which terrorist conversations are transmitted or intercepted.

Some narratives use the trope of transmission to relay an embodied sense of the network. In "Doubles for Darwin" (Pat Sullivan, United States, 1924), Felix the Cat attempts to answer a newspaper advertisement by the Evolution Society to prove that humans are descended from monkeys. To do so, Felix visits the Trans-Atlantic Cable building and tricks the attendant to gain entry. In the cable room, he pulls a large lever like that of a slot machine, and locations flash up on the screen. Each signifies an extreme point in the Atlantic Ocean: Chile, the North Pole, and South Africa. Counter to the traditional Atlantic Cable narrative, the cable here is not signified as an Anglo-American route but is visually translated as a fantasy of global travel: one could randomly transport oneself anywhere in the world, even to places where there is no literal cable route. Felix settles on South Africa, types up a telegram, and wriggles his way through the cable under the ocean. The sea is not superseded in this extension of mobility but becomes a significant force in the story. Felix is pursued by other underwater inhabitants that attempt to cut the cable to stop him. When they do, Felix emerges from the circuit, appearing to transform from the material of the cable itself (figure 2.6). Felix becomes the signal, and following him, we get a firsthand glimpse of its experiences as it moves through the environment. The comedy of the cartoon is oriented around a series of reversals, in which a fish fishes for Felix, a bird chases him, a monkey acts surprised at a human's amusing actions, and finally Felix asks South African monkeys if they are ancestors of humans—when it is suggested that the opposite is true, and monkeys "come after us." In each of these instances, equalization between different actors does not occur, and humor is instead generated from the inversions of hierarchy. The film visualizes and ridicules such aspirations for equalizing interconnec-

FIGURE 2.6. Felix is transformed from the material cable.

tions, links cable transmission with the comedy of reversals of power, and, as it reveals the environments through which cables extend, depicts a new way for us to imaginatively inhabit the network.

Like the *Felix* cartoon, Renny Nisbet's sculpture "Transmission" attempts to foster an embodied sense of the process of signal movement. Located in a garden at the entrance to the Porthcurno Telegraph Museum in Cornwall, England, "Transmission" consists of a series of short posts structured into a maze for visitors to navigate using Morse code. Describing the project, Nisbet says that he aimed to give his audience a sense of the process that electrons went through as they traversed a circuit. Rather than immobilize and abstract telegraph technologies, "Transmission" encourages its user—like Felix—to become waves; to encounter, sense, and even play an infrastructure.[52] In practice, the sculpture works imperfectly. As they follow the sculpture's suggestions, participants are not channeled forcefully or directly between posts. They can easily step out of the maze, become distracted by other artwork, or wander over to the nearby tennis courts. If there is a story conveyed by "Transmission," it is one about the practical realities of electrical transmission in the absence of insulation: signals (and people) without walls, barriers, or guides wander and diffuse into their surrounding environments.

Narratives of transmission often take the form of a travel narrative in which a protagonist tracks a signal to document its diverse environments. Perhaps the best-known popular representation of undersea fiber-optic cables is Neal Stephenson's "Mother Earth, Mother Board," written for *Wired* in 1996. Stephenson follows the laying of the Fiber-Optic Link Around the Globe (FLAG) cable to several of its locations between Europe and Japan. He describes the trip as a "new field of human endeavor called hacker tourism: travel to exotic locations in search of sights and sensations that only would be of interest to a geek."[53] Hacker tourism is a form of cultural tourism—Stephenson spends half of his time commenting on the technological features of fiber-optic cables and half of the time commenting on the distinct cultural features of his surroundings. In this regard, his story can be traced back to cable-ship narratives, such as Florence Kimball Russell's *A Woman's Journey through the Philippines on a Cable Ship That Linked Together the Strange Lands Seen en Route*—these resemble sailors' logs and are part discussions of the cable and part travelogue.[54]

Although Stephenson's narrative focuses on a cable that is yet to be established, it breaks from the typical structure of the connection narrative by directing the reader's attention to events that will occur after the moment of the cable's initiation. At the outset Stephenson seems to fall in line with a broader set of connection discourses proclaiming the end of space, stating that "wires warp cyberspace in the same way wormholes warp physical space: the two points at opposite ends of a wire are, for informational purposes, the same point, even if they are on opposite sides of the planet."[55] This statement is far from what he actually shows in his story, as he quickly turns to the network's limits:

> Netheads have heard so much puffery about the robust nature of the Internet and its amazing ability to route around obstacles that they frequently have a grossly inflated conception of how many routes packets can take between continents and how much bandwidth those routes can carry. As of this writing, I have learned that nearly the entire state of Minnesota was recently cut off from the Internet for 13 hours because it had only one primary connection to the global Net, and that link went down. If Minnesota, of all places, is so vulnerable one can imagine how tenuous many international links must be.[56]

Stephenson travels to Thailand and chronicles how whole villages must be assembled to dig cable ditches stretching across the isthmus that separates the Andaman Sea and the Gulf of Thailand. He discusses the political history that means the cable cannot be routed around Singapore and traces the Japanese training that has influenced decisions about its endpoints. In describing seemingly insignificant spaces along the cable route, Stephenson introduces us to a geography of cable maintenance that rarely surfaces in narratives of connection, from cable companies' island tax havens to manholes that permit access to the network. The significance of the earth itself and its material forms is emphasized at the end of the article: "If the network is The Computer, then its motherboard is the crust of Planet Earth. This may be the single biggest drag on the growth of The Computer, because Mother Earth was not designed to be a motherboard. There is too much water and not enough dirt."[57]

Stephenson's cable tourism has the potential, as Henry Jenkins and Mary Fuller suggest in their essay on spatial stories, to shift our attention away from the narratives of heroic characters and toward spatial exploration, the construction of difference, and technologies' role in these processes.[58] Stephenson does not look for a single source of causality in the cable's development—a feature of almost all connection narratives—nor does he position a single company as a guiding force. Rather, he suggests that the cable is a product of many different cultural practices and methods of work, and as such does not merely pass over the globe but is affected by various cultures on its way. To promote a sense of spatial exploration, he provides a map of his travels by giving GPS coordinates for many of the locations he reaches, "in case other hacker tourists would like to leap over the same rustic gates or get rained on at the same beaches."[59] More critical to Stephenson than encouraging route duplication is the stimulation of infrastructural literacy, which might open up new technological spaces for exploration.

Ultimately, the world Stephenson paints is not one that is homogenized, where points at the end of a cable become "the same," but one in which cultural difference constitutes our global technologies. Stephenson writes that during the journey he acquainted himself "with the customs and dialects of the exotic Manhole Villagers of Thailand, the U-Turn Tunnelers of the Nile Delta, the Cable Nomads of Lan tao Island, the Slack Control Wizards of Chelmsford, the Subterranean Ex-Telegraphers of Cornwall, and other previously unknown and unchronicled folk."[60] He describes each of these groups in terms of their contribution to the global network, using the cable itself as an ethnographic device. Stephenson's focus on exoticism parallels the interest in exoticism that was fundamental to the earlier cable-ship narratives, unsurprising perhaps given that the people who traverse cabled geographies then and now have often been white men. "Mother Earth, Mother Board" politicizes the environments of cable development, but the function of the cable for the local people, its imbrications with race and gender, and any questioning of disparities in power is left out of the discussion. The end of the narrative suggests that even though the physical challenges of the planet will become easier, the "one challenge that will then stand in the way of The Computer will be the cultural barriers that have always hindered cooperation between different peoples."[61] Stephenson suggests that hackers with ambitions to work in cable technology must be able to navigate and surpass these cultural differences as much as the physical ones.

Blum's *Tubes* consciously follows Stephenson's essay in its infrastructural tourism.[62] According to him, "Every word written about this topic [undersea cables] owes a debt to Neal Stephenson's epic 1996 article."[63] In *Tubes*, Blum travels to the locations of the physical Internet, seeking to find its monuments and most important sites to "stitch together two halves of a broken world—to put the physical and virtual back in the same place."[64] Like Stephenson, Blum focuses on the interfaces between Internet infrastructure and the earth's crust. He visits the "spiritual home" of undersea cables in Porthcurno (which he refers to as "the cable world's Oxford or Cambridge") and the stations where they terminate.[65] And like Stephenson, he concludes that the Internet "wasn't a physical world or a virtual world, but a human world. The Internet's physical infrastructure has many centers but from a certain vantage point there is really only one: You. Me. The lowercase i. Wherever I am, and wherever you are."[66] By tracking infrastructure, Blum rewrites people into these networks: without builders, operators, or users, there would be no cable infrastructure. Although narratives of transmission follow cable technologies, they almost always do so to reflect on the human dimensions and embodied experiences of these systems.

Narrating Operations

Disseminating information at the speed of light around the world, undersea cables contort our understanding of time and space. The Internet has led to innovations in narrative form and has become both subject and medium for digital writings and rewritings by its users, who often present it as a dispersed and rhizomatic network. The discursive frames that have been used to narrate the undersea network that supports the Internet, however, have remained limited, holding surprisingly close to a classical narrative model. Narratives of connection and disruption follow traditional linear patterns, and both end at the moment the cable is initiated. They articulate dominant ideologies and underscore industrial calls for investment in network infrastructures. Narrating the transcendence of material worlds, stories about connection and disruption are discursive strategies of insulation that obscure operational cables from view and appear to keep power flowing without interruption through the cable system.

Given that cables are largely hidden and secured, in part due to their status as "critical infrastructure" and their significance for national security, public discourse about them becomes all the more important in shaping our relationship to transnational Internet infrastructure and our opinions about the public policies that govern it. If we believe that the important work of cables is simply in laying them, we are less likely to see the amount of money and effort that goes into sustaining the network and the industry, and we may not properly plan for its continuity. These narratives can also affect where we locate the origin of any intended political change. If moments of disruption are the only key narrative events, then we might miss the more nuanced ways we could participate in cable development, such as in the construction of new modes of user engagement and the reimagination of technological politics.

To provide a counterweight to connection and disruption, I have highlighted a set of discourses that short-circuit traditional modes of representing cable infrastructures. Drawn together in this chapter, they form a scaffolding for the creation of new cable narratives, especially ones that focus on operational infrastructure. They include novels such as Griesemer's *Signal & Noise*, which revise heroic connection narratives about the Atlantic Cable; articles such as the one in the *Economist* that point to the regularity of cable breaks and criticize our eagerness to accept terrorism as an explanation; images such as Taryn Simon's photograph, which direct our attention to what is missing in the frame; and animated works such as the cartoon with Felix the Cat that depict an embodied experience of signal traffic. These expand the static spatial and temporal parameters governing what events can be considered cable history. They reveal how a diversity of actors and activities, from fishermen to cable maintenance, are important to today's networks, and the ways that cultural differences are, and should be, constitutive of global infrastructures rather than subordinated by them. The rest of this book instantiates this approach. Each following chapter tells a story about the history of a different kind of node: the literal nodes of the cable station, the pressure points of cable landings, the islands that mediate network traffic, and the ocean environment itself. Following signals through these zones, this book is structured as a transmission narrative—it attempts to connect virtual worlds with their surrounding social and natural environments and reveal, in Stephenson's terms, the importance of material and cultural "drag" to the growth of networks.

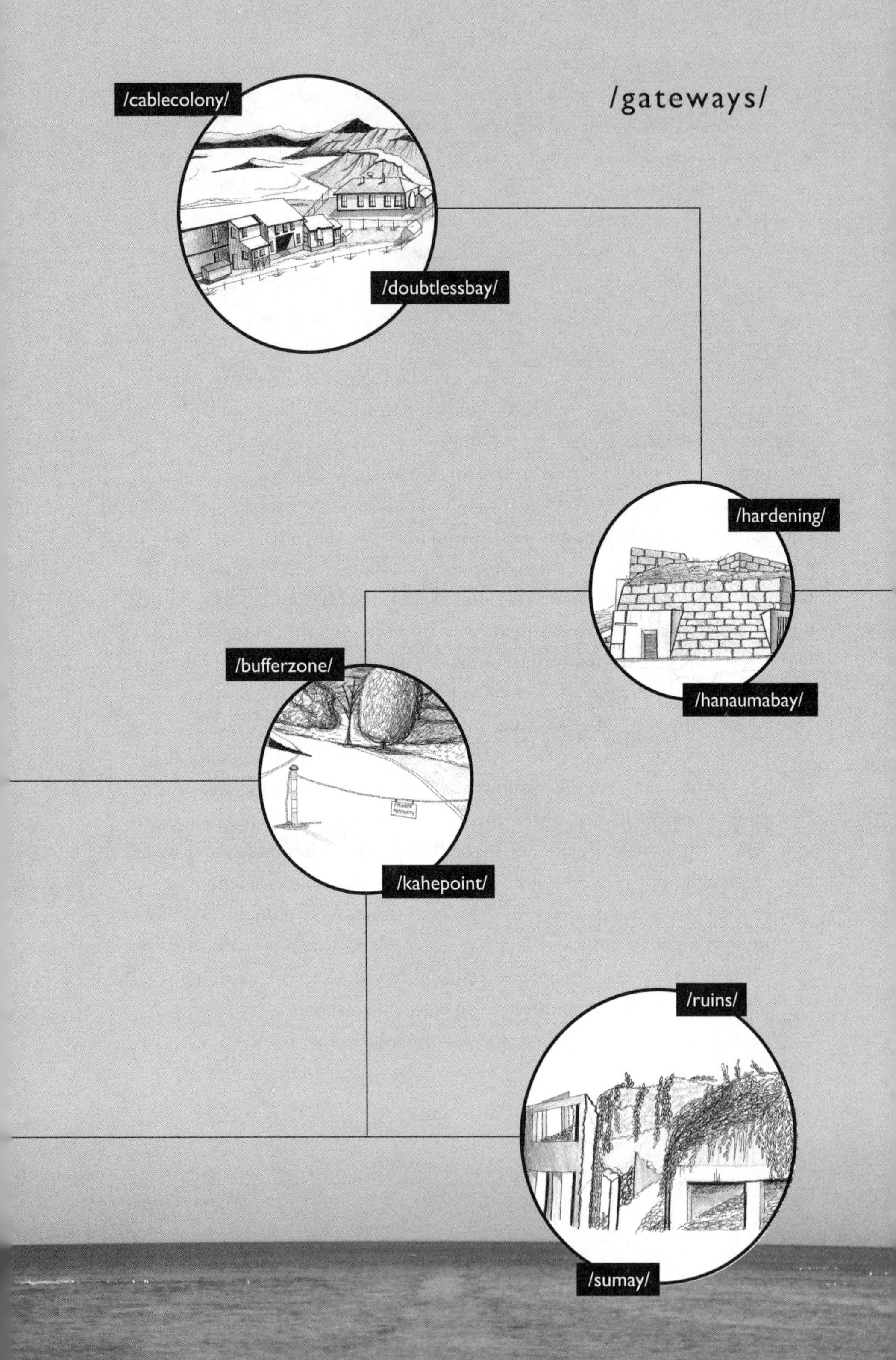
/gateways/
/cablecolony/
/doubtlessbay/
/hardening/
/hanaumabay/
/bufferzone/
/kahepoint/
/ruins/
/sumay/

THREE

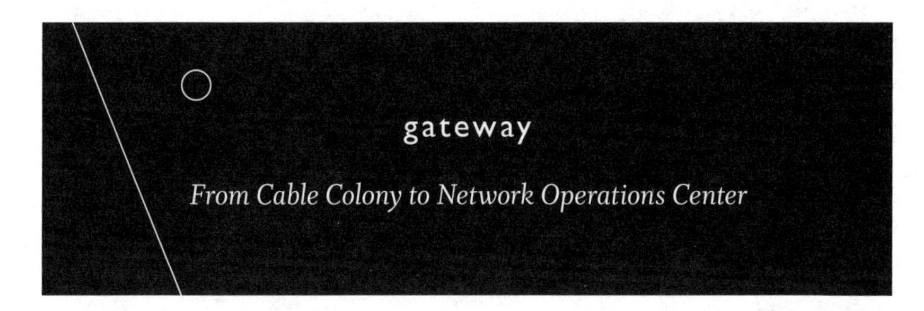

gateway

From Cable Colony to Network Operations Center

In 1962, during a period of technological and political transition, the Keawaʻula cable station was built on Oʻahu's west shore for the landing of the Commonwealth Pacific Cable (COMPAC) (figure 3.1). As telephone cables extended across the oceans, the geopolitical balance of telecommunications power was shifting from the British to the Americans. Although the All-Red Line had landed at British-owned Fanning Island, as discussed in chapter 1, the COMPAC system landed on United States territory. Keawaʻula interconnected historically separate British and American systems—a climax to a narrative of connection, but a starting point for a history of the cable station. Given the perceived threat of nuclear war, territorial placement was no longer a sufficient form of protection, and Keawaʻula was located underground in a fallout shelter. Though run by a Canadian company, the station was constructed according to U.S. military specifications, with walls between eighteen and twenty-four inches thick, showers where employees could wash off radioactive material in case of a nuclear attack, and kitchens stocked with enough food for thirty days. These

FIGURE 3.1. Keawa'ula cable station, O'ahu, Hawai'i.

forms of protection were strategies of insulation designed to shelter signal traffic from potential turbulence during the Cold War.

Visiting the station in 2009, I realize that this gateway has been built not only in relation to changing geopolitical regimes, but also in relation to local social practices. When I approach the building (after driving past the checkpoint for the U.S. Air Force satellite tracking station), I see an elegant set of steps leading up to a door, a front wall composed of square glass blocks, and a long ramp on the right-hand side that makes it wheelchair accessible. Designed by the Honolulu-based firm Martin & Chock, Keawa'ula's aboveground expansion in the early 1980s won several architecture and design awards. One cable engineer described the new station as a palace.[1] Inside the front door, there is a small couch for visitors, a receptionist's desk, and locally inspired artworks, including a tapestry whose caption claims that it "reveal[s] the feeling of Hawaiian water . . . a vital vehicle for communication for the ancient Hawaiians just as it is today." The station even has ocean views. Its manager tells me that the expansion was influenced by a visit from the CEO of Teleglobe Canada, the company that operated the station at the end of COMPAC's life. With little space for gathering inside, the twenty staff members held their meeting in a small and cramped lunchroom. Afterward, the CEO came outside and saw the station workers joking around and hanging out in the trees like monkeys. The manager comments that they must have made an impression on the CEO, and—partly inspired by the cable community—Teleglobe built a new, staff-friendly environment that would support local labor through innovative architecture. The redesigned station grounded the communications hub in the connective capacity of the Hawaiian environment while leaving intact its Cold War foundations, which continue to influence the architectures of the future.

There is no receptionist in the lobby to greet me when I arrive, though I am being watched on cameras scattered throughout the complex. Only two people have clocked in at that point, and the station manager himself comes to meet me at the front door. Far from the bustling community of the early telegraph days in Honolulu or the telephone cable era of the 1960s, Keawa'ula is almost empty—half of its technicians were cut in the downsizing of the early 1990s. Much of the labor has been outsourced to network operations centers or delegated to computers. The manager's time is spent monitoring computer systems, conducting negotiations about where signals can be routed, and dealing with local communities. The station's large conference room, complete with a long center table and several high-backed chairs, appears as if it has not been used in years. AT&T is considering changing it into a storage room. After descending three flights of metal steps in a large concrete stairwell, I am shown the still-functioning battery plants that power the equipment and multiple backup generators. Everything is redundant except for the workers. The manager points out that a dwindling labor force might be the weakest point of cable networks today.

This chapter documents the changing cultural geographies of Pacific cable stations and describes how the spatial practices of operation, upkeep, and labor, together with the broader histories of colonization and militarization, have affected the gateways to our undersea networks. Cable stations are the buildings where cables terminate after they come ashore at a cable landing point. Here, transmitted signals—including dots and dashes, voices, and data—are routed from international cables to national backbones or other international segments of a network, or to domestic networks operated by local backhaul carriers. Like seaports and airports, cable stations are gateways, nodes that function as a region's entry or exit point, house a range of technologies, and comply with standards that knit together heterogeneous communities of practice.[2] At a network's gateway, different forms of traffic are interconnected, and reciprocity is established between these diverse circulations. Transportation routes are leveraged to support labor, biopower is channeled to transfer encoded information, and electrical grids are harnessed to transmit signals miles under the ocean. Christian Sandvig has suggested that by studying such gateways, we can better understand the form and boundaries of an infrastructure.[3]

Gateways are strategic geopolitical locations; centers of power where messages can be delayed, censored, or intercepted; and sites of potential disruption, where interference can be produced in a multitude of ways (from shutting off power to introducing a bug into the system). Strategies of insulation designed to protect the cable's electrical and political currents have been the most critical at transoceanic cable stations. At the same time, cable stations are the inhabited architecture of international networks: they have been tightly knit social spaces, places constructed not only by technological needs but also by the work of a cable community. Even today human labor and embodied experience remain integral to the maintenance of global information exchange. As sites of transfer and transduction, cable stations are the places where access must be strictly regulated and governed, boundaries are deliberated and defined, and the inside and outside of the network are determined.

The spatial organization of the cable station fluctuates with shifts in its surrounding environments. As the design and architecture of Keawa'ula testify, from its thick concrete walls to its surveillance cameras, the cable station has been crafted in relation to anticipated geopolitical interference, including both historical disruptions and dominant cultural fears. In the telegraph era, the station took the form of a "cable colony" that had to be insulated from a threatening natural and cultural frontier, although it was dependent on these environments for support. During this period, the network's boundaries were modulated through a regulation of the body of the cable worker. In the coaxial era, the boundary of the network shifted to the walls of the station itself, and the strategies of insulation were redefined in terms of access to station architectures. Companies constructed the cable station as a "closed world" while simultaneously reaching out to local populations during a period of decolonization.[4] Today the development of fiber-optic cable stations has shifted toward making security profitable and the environment of the station visualizable. Strategies of insulation are inevitably partial: what protects global communications nodes and routes in one historical context might make them vulnerable in another. Despite all attempts to construct the station as an autonomous place, strategies of interconnection are inevitably needed to ground the network in existing spatial practices and infrastructures, including a capable workforce; regional transportation networks; and water, power, and sewage systems. As this chapter moves through a series of node-centered narratives about the station's cultural history, it delineates how the perceived boundaries of the cable station have shifted—from body to architecture to knowledge—and, by shaping contours in cabled environments, have anchored the undersea network in these sites.

Telegraph: The Cable Colony

Whether it's war or / It's peace we declare, / The cables are nerves, / But the men must be there / To tend them and mend them / For that is their share— / Without fuss!—H. F. B, "Cablemen"

As described in chapter 1, early global telegraph cable systems often conformed to the geography of colonial empires, and, rather than seeing the ocean as a threat, they used it to insulate the cable from social conflict. Cable stations were the vibrant and visible nodes of these networks. In marked contrast to the secrecy that surrounds the station today, cable stations and landing sites were depicted on cable envelopes and commemorative postage stamps, and at times their addresses were even indicated on cablegrams delivered from them. Some were elaborately built complexes that constituted the central structures of the area. The multistory cable station in Bamfield, Canada, with its sweeping hilltop view of the Barkley Sound, was designed by a leading architect in nearby Victoria, Francis Rattenbury. R. Bruce Scott, a cableman who worked there, compared it to a French chateau: it had close to fifty rooms, including not only telegraph offices but also accommodations for servants and cablemen, a billiard room, a music room, and a library with books about the British Empire (jokingly referred to as the "All Red Room").[5] The station bathrooms were equipped with electric bells with which one could "summon a Chinese servant when in difficulties of some sort or other."[6] There were a number of separate houses on the property for married men. Cable stations also had to be built to protect against natural phenomena. Guam's station was "self-contained": it was intended to be fire, earthquake, and typhoon proof; had its own water supply and sewage and refrigeration systems (including an ice plant); and was lit using acetylene gas manufactured on site.[7] Described as a "cable colony," the station was a self-sufficient and insular space where companies reproduced a microcosm of empire.

To ensure that the social composition of stations remained distinct from that of the local community, cable companies engaged in a range of strategies of insulation. At first, they tightly controlled the national origin or affiliation of the cablemen. Stations operated by British companies were initially staffed only with British workers. On the Commercial Pacific telegraph route, a cableman remembered that station employees had to declare their intention to become American citizens and that, fortunately, "this order has gained citizens for Uncle Sam."[8] In early cable stations, local men were not trained as operators, despite the fact that men from abroad were not necessarily any more qualified.[9] Recruiting workers at a young age (as early as fifteen), the companies socialized cablemen into the culture of the cable system, where they would spend much of their lives.[10] The men were required to work long hours on the line and to shift from one station to another with relative frequency. On-site accommodation meant that cablemen would not have to find homes in the local town (even when they were available). These policies ensured that cablemen would maintain connections to a distant homeland and other cablemen instead of to local residents. The policies also helped create a relatively insular cable community—a social structure that kept the men from becoming attached to specific locations; stabilized flows within the network; and prevented information, expertise, or resources from diffusing to individuals outside of the cable colony.[11]

In some places the station's social insularity, combined with the cablemen's technical knowledge, led to improvements in cable technology and practice that eventually boosted the connectivity of the emerging cable network.[12] The production of the cable station as a completely insular node was also potentially dangerous: it could result in disruptive practices and generate interference with system operations. In many stations, cablemen drank extensively to cope with boredom in harsh and remote environments, and such drinking could make them sick and cause them to leave duty early.[13] Station managers discussed and combated this problem. As it introduced new social stratifications into communities, the cable station could also negatively affect social life. In his study of Heart's Content, Ted Rowe observes how the transatlantic cable landings resulted in the global dissemination of news about the local people's backwardness, the shabbiness of the houses, and the unpleasant odor of drying codfish.[14] This characterization ignored the "tightly knit, resilient culture" of Heart's Content and offended local people.[15] In other places, less amicable relations between station and indigenous people of the area structured community life. Such was the case in Darwin, Australia, where Aborigines were accused of stealing things from the station.[16] The insularity of the station could generate resentment and inequality, creating an imbalance in the local culture.

Completely insular nodes were difficult and expensive to maintain. Cable companies had to pay salaries and transportation costs of administrative and technical staff, including engineers to maintain the system. In 1907–8 the Pacific Cable Board reported spending £38,764 for station salaries and expenses, a figure equivalent to millions of dollars today.[17] Companies needed to fight the effects of storms and other forms of environmental degradation. They also needed to ensure that power and water systems and other necessary infrastructures were available for employees, including medical support (telegraphers faced diseases such as yellow fever in the tropics). In many cases, complete insulation was impracticable, leaving employees subject to frictional forces that might disrupt their work. In the early days of Guam's station, cablemen faced difficulties in obtaining water (it had to be carted in from another town) and problems with tropical insects, and they were not allowed to purchase fresh meat or ice from government stores.[18] Many station employees got sick. Cablemen in Newfoundland had to cope with temperatures below freezing. In Darwin white ants consumed the insides of walls. On Midway Island there was almost no vegetation when the cablemen arrived, and the white sand created a blinding glare every time they stepped outside (the interior of Midway's station was painted green to compensate). In Rio de Janeiro cablemen confronted extraordinary storms. A letter from the superintendent of the station reports that a terrible storm "has rendered the building in most parts dangerous and uninhabitable. . . . There were only three bedrooms dry, all the others were inundated; beds and personal effects were all drenched."[19] Given the station's limited budget, they had barely enough money for silverware. If a remote cable station lacked adequate insulation, its operations were in danger not only because of possible structural damage, but also because the poor living conditions could affect operators' morale and accuracy.

For economic, technological, and social reasons, cable stations could not be completely autonomous places: they relied on and participated in local infrastructure and practices. Most significantly, cable companies depended on local labor. Indigenous people served as personal servants, cooks, gardeners, and butlers. Countless photographs and descriptions recount their help in pulling the cable ashore and completing the circuit of transmission.[20] Cablemen also relied on local infrastructure (including boats, roads, and "baggage coolies") to explore the surrounding areas and transport supplies. If local labor was nonexistent or considered unsuitable, servants and laborers often would be sought from places such as China, Japan, and India. In its review of Rodriguez Island, the Eastern Telegraph Company reported: "It is advisable to obtain Cooks and Butlers from Mauritius as local material [are] unsuited to these positions except in exceptional circumstances. The wages paid to Mauritians is [*sic*] higher but worth it. Other servants obtained locally."[21] Drawing on the resources of existing populations and rerouting circulations of colonized bodies were strategies to locally interconnect station operations, offsetting the turbulence of natural environments as well as potential social conflicts.

Another social practice that facilitated interconnection at the cable station was the development of a culture of leisure. Nested in the wilderness, many Pacific cable stations became centers of social activity. The *Guam News Letter* recorded in 1917:

> Guam was taken by storm, notwithstanding a rising barometer, when the Cable Station Dance Club broke loose in Sumay on August 18th, and all hands hope for repetitions. We are convinced that we have the best Cable Colony in the world here in Guam and we voluntarily undertake to defend that title. . . . The floor was all to the good par exellence [*sic*], the music inspiring, the weather made to order, the grounds were transformed into a Garden of Eden, and the automobiles were so numerous as to necessitate parking the surplus in the vegetable gardens.[22]

Guam's cable station had its own sailing crew, which raced the crew of another town. It was the center of automobiling parties, billiard tournaments, and even the attempted installation of a swimming pool. In Fiji, tennis, rugby, and picnicking were popular pastimes, and several of the men owned boats, which they took out for weekend parties. Visiting other parts of the islands, they brought the chiefs the "customary bottle of whisky" in order to enjoy fishing in their lands.[23] On Fanning Island, the cablemen held canoe races and formed a gun club, at which they shot clay pigeons. Bamfield had very little indoor entertainment until the cable station arrived, but after that the following were established: a Saturday night movie showing that was open to the public, a dance hall that featured a five-piece cableman orchestra, and annual dances at the end of the summer and New Year's (which local residents would travel for hours to attend). "It was like living in a country club," one cableman recounted.[24] Occasionally, stations even became tourist destinations open to the public. During some summers "droves of tourists" would arrive at Bamfield, and the cablemen would host visitors at the station.[25] A summer cruise to the Northland region of New Zealand made a stop at the Doubtless Bay station, and the superintendent greeted tourists on their trip. These activities were developed with the support of superintendents and company managers and at the initiative of cablemen who sought out connections beyond the cable colony.

These events had two functions: solidifying the cablemen's sense of a cable community, as reports of similar events were shared across the network, and serving as a way of connecting with local communities. Like many colonial structures, the cable colony was grounded via its ties with the local elite, the only people who were able to attend many of the events. In turn, operators were often invited to activities hosted by the towns. The events would sometimes result in substantial interpersonal connections, including marriages between cablemen and local women (the Bamfield station was regarded by the Port Alberni girls as "a marriage bureau").[26] In some places, indigenous people were welcome to participate, especially in sporting events. According to a report from Fanning Island in the 1920s, "The various native events occupied the main part of the programme. The first element of excitement was furnished by the native weight putting. There were no casualties, as, early in the day the spectators found that the only comparatively safe place was well behind the competitors. The native women's Tug-of-War was exciting."[27] The cable station even had their own Gilbertese team that would play against other Gilbertese residents of the island. The cablemen's magazine shows cable community children dressed up in Gilbertese attire and observes that they had been taught how to dance by the natives.[28] At Norfolk Island, sports were open to "all comers," and a special event was organized for the boys of the Melanesian Mission, though the cablemen reported that "they were too diffident to compete in the presence of so many 'papalangi' (white people)."[29] In other places, such as Darwin, the cablemen would go hunting with native men, drawing on their knowledge of the environment. The culture of leisure at the cable station was an integral part of cable station life, and these local connections could help to negate any cultural imbalances created by the system's insularity.

Cable stations sculpted contours into the local social topography, often becoming a center of gravity that attracted resources, commerce, and further development. In some places, the resources of the cable station became available to local publics, who thus benefited from cable investment. This could elevate them in stature relative to people living in other areas farther from the station. The station in Guam played a large role in the economic prosperity and development of the nearby village, Sumay, and many villagers worked at the station. On a redundant section of the Pacific Cable, Scott recalls that the man on night duty with nothing to transmit would send newspaper extracts to Norfolk Island "for the local 'newspaper.' These were typed up by the Norfolk Island operator and tacked up on a tree at the crossroads for all to read."[30] This news source was known as the "Tree of Knowledge." Many structures established in Heart's Content to support the cable station, such as a new school, were limited to those who could pay (and were therefore not available to poor local residents); others, such as new churches and a new water supply system, were eventually available to everyone.[31] As time progressed, local men in some places were trained as operators, and cable stations expanded as centers of employment. At stations such as Southport, Australia, operator schools were created on site.

The cable station building was porous, and the cultural mixing in these remote environments meant that social boundaries could also become unclear. Indigenous labor made the station's operation possible but simultaneously created a structure of dependence that produced anxiety in the cablemen. One station worker expressed surprise when he arrived in Darwin to find "white men doing coolies' work" and saw "Chinamen serving white men on almost equal footing."[32] Cablemen expressed an ambivalence toward the local population that was typical of colonial relationships, oscillating between distancing themselves from and proclaiming their affinity for indigenous people. When the Gilbertese at Fanning Island were replaced by Chinese laborers, one cableman wrote: "Although all the 'wives' ever at Fanning Island have had Gilbertese female servants, and often silently cussed them—not too silently, sometimes—they have all had an underlying affection for these hefty, incompetent and, if not constantly watched, dirty servants. . . . Their happy, indolent disposition always responded spontaneously to a few kind words . . . the passing of the Gilbertese is a calamity. Their eternal 'Limwi' (O, bye and bye) has become such an integral part of the life and the climate here that it will never be replaced."[33] In some places, the native islanders were seen as part of the cable community, but their inclusion was structured by a colonial ambivalence that simultaneously deemed them objects of affection, and "dirty," "incompetent," and "indolent."[34] Local populations were necessary to the maintenance of circuits and also constituted the biggest threat to them.

The boundary between the inside and the outside of the network had to be actively maintained—not at the walls of the cable station, but at the body of the cable worker. In his study of Cable & Wireless, Hugh Barty-King reflects on the importance of the cableman: "Eastern Telegraph's fixed assets consisted of many thousands of miles of submerged cable, numerous telegraph stations, and expensive equipment but, most important of all, *people*, who in 1872 began to build up the tradition of loyalty and expertise in the specialized field of Overseas Telegraphy, the preservation of which was always the most telling argument against fragmentation when technical, economic, and political reasons appeared to be overwhelming."[35] Cable transmission at this time was a distinctly human practice. Operators responsible for sending and receiving telegraph messages literally became part of the circuit. One cableman recollects: "During our solitary watches of six hours, we were, for the time, the only links completing communication between Home and the Colonies. . . . We were keeping up that important connection single-handed. . . . Until relieved at the end of a duty, intercourse between Australia and the rest of the world depended entirely upon the man at the circuit."[36] This was an intense job. One cableman argues that even though people believe cable operation is "a humdrum, mechanical occupation," it is in reality "a business of thrills and chills," in which the men are some of the first to bear the transmission of "world-shaking events," from earthquakes to wars.[37] As their work spanned time zones, cablemen's bodies had to accord with global rather than local temporalities, with the men working overnight to receive traffic during business hours in another part of the world. When they stepped into their place of work, cablemen occupied this global space, a zone of speed and pressure.

The mental acuity and competency of the men directly affected the speed and accuracy of transmission. In the early days, the operator's work consisted of watching a light move back and forth across a small strip, a practice that strained the men's eyes. This would take place in a room where much of the light and air was blocked out, and in the tropics work might also be conducted in extraordinary heat. The signals would have to be written out in full and then retransmitted on the connecting circuit (when recorder instruments were developed, the conditions improved, and operators could have more light and air). In training, emphasis was placed on cultivating the correct wrist action so that an operator could endure long hours of work. The signals on undersea communications cables, although in Morse code, were not transmitted via a set of distinct dots and dashes, but instead as minute waves in an uninterrupted line resembling a "seismograph record" (a technological practice that further distinguished the undersea cablemen from regular operators).[38] Decoding the signal required intense interpretive effort, and the ease of reading would depend on the cable's geography. Operators around the Pacific recalled the difficulty in reading the messages between Fanning Island and Bamfield, Canada (the longest cable in the world).[39] Geological and atmospheric phenomena could also create distortion. Overall, cable work depended on and took a toll on the men's bodies: the clarity of the message directly reflected the skill of the operator and his geographic location.

The cable station was a place where microcirculations linked with global currents and local practices could disseminate through the network (from technological developments to excessive drinking). The body of the cable worker —a site through which messages and information passed and where they were interpreted—formed the key gateway to the system and the network's most important pressure point. Very small movements here could have large-scale effects; the cableman's body was understood as the site that was most susceptible to variance or interruption of flow. As Jussi Parikka argues, in networks in the nineteenth and early twentieth centuries, "it was most often *people* who occupied this position of the 'parasite,' or intruder in between of *trans*-mission. . . . Human beings were often perceived to be a guarantee of functionality and security in man-made systems, but it seems that they were also a potential source of noise."[40] The body of the cableman was the place where the border between the network's inside and outside was enforced, the guarantor of quality performance. As a result, operators were held to strict standards of performance. Errors in transmission on the job were tallied and tracked.[41] Younger employees were discouraged from marriage, and insobriety, insubordination, and using "improper" language or "quarrelling on the instrument" were forbidden.[42]

Potential interferences included not only the failure of the cableman's body but also its susceptibility to the social mingling and diffuseness of cultural boundaries at the cable station. A key set of discursive strategies of insulation, circulated via cable station magazines, functioned to regulate bodily performance. Since the beginning of cable work, cablemen have disseminated station occurrences via journals and magazines: from chronicles of cable laying in the *Atlantic Telegraph* of 1865 and the *Great Eastern Telegraph and Test Room Chronicle* of 1866, to the Eastern Telegraph Company's monthly family magazine, the *Zodiac*, launched in March 1906. The latter circulated beyond the stations and even became a geography primer in at least one school due to its extensive documentation of foreign locales.[43] Other examples include the column "Cable Station Chatter" (written from 1915 to 1917) in the U.S. military publication *Guam News Letter*; the *Creek Weekly*, which showcased the "literary and artistic talent" at the Bamfield station; the *Islander*, from Ascension Island; the *Whites Bay Rag*, from New Zealand; and the *Recorder*, published by the Eastern Telegraph Company's Training School. The Midway Island staff produced the *Gooney Clarion*, "A Weekly Newspaper Published Every Now and Then" (an article in its farewell edition records an "exciting day" on which a ship passed nearby the island, but "it is hardly necessary to add, did not stop").[44] Reviewing the *Heart's Content Aurora and New Perlican Trumpet*, a local editor in St. John's, Newfoundland, described it in 1872 as "a safety-valve whereby many escape the exuberant comics of the jolly dogs of Cable Terrace, who otherwise would run very great danger of bursting their respective boilers. [It] is, the public will regret to learn, printed for private circulation only—these Cable Terracemen thus proving themselves to be to some extent aristocrats and exclusives."[45] Even though some of these newspapers and magazines circulated beyond their stations, they were nonetheless shaped by the station's insularity and were sites through which the cablemen could participate in an imagined cable community.

These publications served as a strategy of insulation that upheld the boundaries of the cable network: they both negotiated the anxieties of cablemen about mingling with local people (who were depicted as outside of the system) and reinforced the model of the "correct" bodily performance (training cablemen how to remain inside the system). Some spreads, such as one titled "Children of the Submarine Cable Service" (featuring photographs of cablemen's offspring), encouraged cablemen to view the service as their family.[46] Cable photography more often mediated racial difference: photographs featured bare-chested natives from many locations engaged in "primitive" activities. Extensive narrative descriptions were provided of the natives of different locales, from the "cannibalistic instincts" of Australian Aborigines to the Philipino, "an appropriate present for a deadly enemy."[47] (There was no hesitation in documenting the stations, cable trenches, and cable routes: built architecture was not seen as the site where security would be compromised.)

In many cases, racial difference was used as a point of humor, allowing cablemen to diffuse their anxiety through laughter. In one article about Midway Island, below the names of the Western staff, the author includes the following note: "Also Staff of Chinese servants and Japanese labourers, whose names I beg to be excused from attempting to write down, my Oriental secretary having fled from my sight upon my attempt to engage him in Asiatic lore."[48] Acknowledging the necessity of Japanese and Chinese labor in the station, this cableman makes a joke of their names and ultimately does not include them in the list. A regular item in the *Zodiac*, titled "Savage People We Have Lived Amongst" featured cartoon-like figures of "The Bomb-Throwing Bengali," "The Fighting Sikh," and Australian Aborigines. Another illustration, called "Old Time Staff Group," depicts two fully clothed cablemen on a desert island alongside racialized caricatures of natives (figure 3.2). The image depicts the cablemen in a hostile environment: there are scorpions, a box of quinine (a medication used to treat malaria), and an office that consists of nothing more than a palm tree. Both cablemen look uncomfortable, and one of the black men holds the circuit in his hand, an illustration of the cablemen's reliance on native labor. Despite the apparent vulnerability of the cablemen, the seminude caricatures of the natives enabled readers to conceptualize themselves as superior. Staging and restaging the encounter of racialized difference, the *Zodiac* used humor to undermine the agency of the indigenous people whom their readers depended on.[49] Theatrical shows and performances were put on by the cablemen that reproduced stereotypes of ethnic or national Others and further served to reinforce the boundary between insiders and outsiders (one troupe on the Cocos Islands, named The Boozy Blacks, performed in blackface).[50]

The *Zodiac* portrayed the work of the station as part of the colonial project—to civilize and tame the natives. One series of images, titled "A Civilising

FIGURE 3.2. "Old Time Staff Group." From *Zodiac*, n.d., © Cable & Wireless Communications 2013, by kind permission of Porthcurno Telegraph Museum.

Service—A Study in Evolution" depicts a dark-skinned man in a suit, "Amit," presumably a cable worker, next to an image of a wild native, dressed in nothing but a loincloth, "Amit's Father" (figures 3.3 and 3.4). Here the work of the cablemen and the cable station is overtly depicted as a part of the "civilizing" work of colonial presence: to train indigenous populations in proper British behavior and self-presentation. Although depicting the subservience of the native to the cableman's performance, the images nonetheless point to the fundamental racial difference between Amit and the implied British reader. The cable magazine's representations, as they mediated racial difference, coded the correct performances of the cablemen, eased anxieties about the inevitable connection between them and local people that occurred at the station, and formed a key support infrastructure that helped to reinforce the boundary of the network: the body of the cableman.

In the 1920s a regenerator system was deployed across the undersea network. This reboosted the telegraph signal without the need for any human

I.—AMIT.

FIGURES 3.3 AND 3.4. "The Civilising Service." From *Zodiac*, © Cable & Wireless Communications 2013, by kind permission of Porthcurno Telegraph Museum.

Drawn by Mervyn G. Skipper

II.—AMIT'S FATHER.

translation, effectively replacing human operators with automatic relay systems. In the early days, a message from England to New Zealand and back might be handled thirty-eight times at nineteen stations along the way, but by 1940 a message could be sent direct with no human contact. Cablemen now supervised the machines and fixed problems when they arose. "Automatic scrutineers" monitored the passing signals, and if they deteriorated beyond a certain threshold, "audible and visible warning by means of a bell and a lamp are given."[51] One cableman observed that the shift "had a demoralizing effect on the cable operator. It was not submarine telegraphy anymore—it was electronics. No longer was he the linchpin in the service. All he had to do was scrutinize tape automatically received, looking for mechanical failures before it was automatically retransmitted. It was very depressing."[52] In the *Zodiac*, one cableman wrote a poem about the transition that highlights the move away from the cableman's body. Although the embodied capabilities of the "fellow with dexterous digits" were once admired, after the implementation of the regenerator, "all is peaceful and quiet. The signals are solid and square. / We sit and we sigh as we watch them go by,—(the Watcher must stick to his chair!)."[53] Another operator writes that romance was now gone from the cable service, in contrast to "the old days," when "one knew who was at the other end of the cable by his touch, or—lack of it."[54]

The operation of the cable station continued to depend on the cablemen and was shaped in relation to their social practices. However, the body was no longer the primary gateway to the cable line, a fact that changed the discourses, work, and culture at the cable station. Less labor was needed at the station (there were cuts in personnel of as much as 75 percent), and the geographic dominance of the cable colony weakened in many areas.[55] The culture of leisure also subsided (preeminence in sports was difficult to maintain with fewer men).[56] Alongside this shift, women and indigenous people were given more important roles in the station, and the *Zodiac* began to demonstrate an increased cultural awareness. For example, a 1926 issue (whose publication was simultaneous with the spread of the relay system) reports: "We have received a letter from Mr. M. E. F. Airey, Wellington, New Zealand, in which he informs us that by using the term 'Chinaman' as we do in our December number . . . we are giving very great offense to Chinese, as 'Chinaman' is considered a term of contempt."[57] In response to the letter, the editors pledged to avoid such "anachronistic errors," as it "is a dangerous thing meddling with anyone's nationality and confusing their terminations."[58]

From the early telegraph era through the development of the automatic retransmitter, the cultural geography of cable stations in the Pacific was shaped by the extension of colonial empires, a fact that was true not only for undersea systems but for many terrestrial networks as well. The cable station became the critical geopolitical node for transoceanic traffic: it was a cable colony intended to be physically self-sufficient and culturally insular, autonomous from its surrounding geography. In reality, British, American, German, and Dutch expatriates came into contact with rural communities, and friction was generated with existing social, natural, and biological circulations. The cable network introduced a set of currents that shifted existing ways of life and that, in turn, could produce interference for cable circuits. At the same time, the global network was grounded via a culture of leisure in local publics and practices, and this kept global flows from becoming imbalanced. However, these interconnections made permeable the boundary between the telegraph station and local geographies. The bodies of cablemen formed a vulnerable gateway at which the network needed to be protected: their bodily performance was critical to the maintenance of global flows. Discursive strategies of insulation—including the racialization of indigenous workers in cable newspapers—helped strategically uphold social hierarchies and keep the cablemen an elite, distinct group, insulating the flows of empire. These practices also produced an imagined cable community, a social infrastructure that came to form a key part of the operation of cable networks. Even after the relay system was implemented and beyond the disconnection of the telegraph cables, the spaces that had been created—the cable community, the built cable station, transport infrastructure, water supply systems, local cultural practices, and natural phenomena (and modes of buffering them)—lingered on and shaped the form of subsequent circulations and, as shown in chapter 1, often became the rationale for keeping routes in the same locations.

Coaxial: Hardened Architectures

As the transoceanic telephone network was extended in the late 1950s, a new set of cable stations were constructed around the world's oceans. Coaxial systems had more extensive technical and architectural requirements, including substantial terminal equipment, and in most places new cable stations had to be built (in some sites, such as Guam, existing stations had been destroyed during World War II). During this period, the geography, architecture, and social practices of the cable station underwent a drastic reconfiguration in light of Cold War concerns. The cable station in the Cold War, like many telecommunications installations at the time, was shaped by what Paul Edwards terms "closed-world discourse": the determination of a radically bounded area, a unified place defined by a central struggle.[59] Although in the colonial era the cable station had been a porous space, in the 1950s and 1960s a set of strategies of insulation termed "hardening" refocused efforts on the architecture of the cable station, constructing it as a closed and autonomous world. Cablemen remained critical to station management and equipment supervision, but the primary site of network regulation shifted from their bodily performance to station architecture.

Hardening was a process of physically fortifying the cable station and often entailed locating it underground in a nuclear fallout shelter. This corresponded to a broader shift in the perception of how power would be intercepted and diffused: the focus was on a potential attack from above. The cable station had become a militarized site during World War II, when there had been numerous attacks on cable stations (including those at Fanning Island and the Cocos Islands). During the war, the cable station at Porthcurno (the "nerve center" of the empire) was moved into a hillside to protect it from attack.[60] After the war, many Pacific cable stations were also relocated underground, including Tanguisson Beach (Guam), Keawa'ula and Makaha (Hawai'i), and San Luis Obispo (California). Philip Kelly, a telecommunications engineer, recounts that security after the war "dictated" that the Oban cable station in the United Kingdom should be built into a hillside, and that in London the system terminal should be constructed in an unused tunnel of the London Underground.[61] In some places, such as Hanauma Bay, Hawai'i, the outside of the station was disguised to look like part of the landscape in case of enemy attack (figure 3.5). Although in the telegraph era, the cables landed in a cable hut on the beach before extending to the station, this was relocated underground in a manhole. Cables that extended across the sand right next to the station could be easily disrupted (even by an axe, never mind a nuclear bomb), but very little consideration was given to protecting the cable's route, manholes, or beach landing point. One report on station security admitted that "there are a number of such manholes along [the] cable route and that in addition, [the] route of [the] cable cannot be concealed and we cannot afford absolute protection in this respect."[62] The only sites at which companies were interested in affording "absolute protection" were within the fences of the cable station.

The American installations in the Pacific, run by AT&T, consisted of a wave of "superstations," as some transmission engineers describe them. They housed an enormous amount of equipment and were built to withstand immense force. The Keawa'ula station was required to withstand overpressures of at least fifty pounds per square inch, equivalent to the force of a wind of over 900 miles per hour.[63] As part of the agreement to establish the station in Hawai'i, the U.S.

FIGURE 3.5. Hanauma Bay cable station, Oʻahu, Hawaiʻi.

Federal Communications Commission required that the Commonwealth "at all times comply with any requirements of appropriate United States government authorities regarding the location and concealment of the cable buildings and apparatus with a view to protecting and safeguarding the cable from injury or destruction by enemies of the United States of America."[64] The Tanguisson Beach station, like the Keawaʻula station, resembled an underground bunker. To enter, cable workers had to descend several stories down a fortified stairwell; to access the equipment room, they then had to go through a heavy metal door. Immediately inside the entrance was a decontamination chamber, a precaution for nuclear attack. Everything inside was designed to be redundant in case the domestic infrastructure on which the station normally relied failed. It was staffed by at least two employees at all times, but it was often occupied by many more. There were completely separated and redundant backup generators to power the building and air-conditioning units, as well as independent battery plants to sustain the equipment.

As they were often located in rural areas, superstations, like cable colonies, had to maintain their own support infrastructure. The station at Keawaʻula was so far removed from the water infrastructure of the Waianae coast that the station's construction involved digging two wells (the primary well and a backup) over two hundred feet down into the ground. As with the earlier Pacific telegraph stations, the Cold War cable station was meant to be self-sustaining: each had a fully stocked kitchen inside. A whole set of "catastrophe spares" for parts of the cable and terminal equipment were stored on site. Inside the building, workers maintained strict temperature and air-quality levels to preserve the "delicate" equipment that "could easily be damaged by humidity and dust-laden air."[65] As a result, in one station's construction "all ledges and projections where dust might lodge were reduced to a minimum."[66] Hardening never did protect circuits from an actual nuclear attack, but it did secure the station from the natural environment, which was especially important in places such as Guam, where typhoon winds reached over 150 miles per hour. When power went out on the island, employees and their families occasionally took shelter inside the station. By constructing the building as a closed world, cable companies were able to stabilize the flow of signals across the Pacific.

As the boundary to be regulated moved from the body of the cableman to the architecture of the cable station, crossing into the network was coded less in terms of performance than in terms of access. Constructing layers of access had become an important concern due in part to the increasing interconnection of different networks at the same station during the coaxial period, as well as the extension of supervisory and control systems. For example, the 1984 Australia–New Zealand–Canada (ANZCAN) cable was equipped with an Automatic Transmission Measuring System, which allowed technicians to remotely supervise repeaters, power feed voltage, current, and even operations at another cable station.[67] As this system could be operated from each station, one could potentially access more than just a single node when entering the building. Nonetheless, there were no centralized maintenance and control centers, and despite the automation of many tasks, embodied human labor remained important to each station's daily operations. A significant staff of technicians would be required to perform routine maintenance on the systems, including testing circuits, coordinating network traffic, and monitoring the extensive terminal equipment. Stations were often manned twenty-four hours a day and seven days a week.

In the colonial era, the station was porous, but in the Cold War era, people who were not part of the network were rarely allowed inside. Concerns about regulating access and protecting circuits permeated discussions of the period. For example, in their survey of a New Zealand station, the Cable Management Committee reflected that the "security aspects [were] fully taken into account": the main equipment would be under constant supervision and the building would be manned continuously.[68] All permanent and regularly visiting staff members would have to obtain security clearances with the police and with New Zealand's Justice Department. Visitors were permitted entrance, but only when accompanied by officers who had been cleared. Cablemen were regularly reminded that the work done in the station was not to be shared with people who could not enter. In one of the stations the cablemen's operations desk directly faced a poster titled "The Official Secrets Act Affects You," which listed the important reasons to keep their activity secret. Secrecy was a strategy of insulation that functioned primarily to obscure the inside of the cable station from vision. One author remarks that "growing up in Heart's Content in the 1950s, unless your father worked at the cable station you probably never got to see inside the door"—a significant change from the early telegraph stations, which accommodated tourists and encouraged public participation in station events.[69]

The politics of access permeated the interior of the station, with sections often subdivided for access to different circuits. In designing the cable station routing for a transatlantic link, Kelly remembers that the Canadians insisted that their circuits be separated at all times from those to the United States; locked rooms were constructed in which only authorized personnel could access the Canadian lines.[70] Local telecommunications companies whose cables interconnected at the station were not allowed direct access to these circuits. This mirrored the cultural geography of routing, which at the time focused attention on the circuits themselves. Kelly suggests that the practicalities of these systems did not always achieve the level of security they aspired to: the "locked" rooms in the Newfoundland station were never really locked in practice.

As in the colonial period, an imagined cable community continued to serve as a social infrastructure that grounded and stabilized traffic flow. The community was produced through the architectural insularity of the station, the shared knowledge of national and technical secrets, the continued dissemination of cable magazines (which, especially in the publications of Cable & Wireless, continued to present the company as a family), and the shared travels of cablemen to training schools and stations across their network. In the large committees of this period that oversaw the construction and operation of cables (the club system), employees from various telecommunications companies would fly to meetings in each of their respective countries. One cable engineer recounts his life after starting work on a new cable system: "It was just 24/7 for the next three years. Hell for my wife, because I was away from home for an extended time. We'd have a meeting in Wellington; we'd have many meetings in Sydney; we had meetings in Canada. Some cable was being manufactured in Japan, so we had to go and see what was going on in Japan. You got to see places you'd have never seen."[71] In my interviews with cable workers who started during the coaxial era, many of them insisted that despite the stress of travel, the excitement of working with people in other countries was part of what had interested them in the job. Through these trips, cable workers developed a shared knowledge of the network's geography—routes, stations, and landings—that bound them together. One management committee observed that the value of these meetings "was not always appreciated by those who had the financial responsibility to approve them," but that there was a "tremendous advantage" to the social relationships that were formed and the exchange of informal knowledge that occurred there.[72] Cablemen of this period continued to imagine themselves as part of a community that spanned national borders and was solidified through their shared access to the network's nodes.

The importance of this labor force ensured the station's continued dependence on connections with local communities and infrastructure. John Cavalli reflects on the development of the Manchester cable station in California, constructed in 1956 and still in operation today: "Establishing this station changed the lives of many local people. Farmers, construction workers, and hired hands were locally employed to pour foundations and to pull the transmit and receive cables up onto shore."[73] Some of the station technicians were hired locally and subsequently married women from the area. Cavalli writes that since their station was so rich in history, technicians created an archival museum to collect its artifacts, including early black-and-white photos and objects such as safety eyewear. The need for interconnection tempered how remote the stations could be. For example, although a cable management committee considered establishing a cable station at La Perouse, they determined that the site would not work well "from the point of view of staffing" because public transportation to the area was so limited.[74] When stations in more remote locales were established, such as the AT&T and Australian Overseas Telecommunication Corporation stations on Guam, special housing had to be built and provided for employees. AT&T hired a local company, Frank D. Perez & Brothers, to build its workers' housing on Guam in Perezville (AT&T also hired local contractors to build its station). The use of local labor, coordinated social events, and intermarriage continued to ground cable workers in the local contexts, even if these events occurred less frequently over time.

During this period, in cable stations across the Pacific, the social landscape was changing in another dramatic way, one that posed a threat from below rather than from above. Many cable stations were located in former colonies that either had recently gained independence or would gain it during the coaxial era. In this context, the network's gateways were redefined: instead of gateways to a colonial empire, they were now gateways to new nations. The cable station in Fiji, once a gateway to the British Empire, was now a site where Fijian traffic interfaced with traffic from other Pacific islands, Australia, Canada, and Britain. At the same time, it remained a node in a system that was predominantly owned and operated by other countries. In some places, the territory on which the cable station was situated belonged to another nation: Tumon Bay was owned by the Australians (and, rumor has it, became the site for the Australian consulate). The Keawa'ula station was Canadian territory and flew the Canadian flag.[75] As conflicts over decolonization emerged across the world, currents extending through the cable station could be figured as a continuing intrusion of a colonial or national power. Stories about communications conflicts—such as a terrorist attack on the Cable & Wireless office in Haifa, Israel, in 1947; anti-British rioting in Indonesia that cut communications to Malaysia and forced staff members to move to Singapore; and plane hijackings in Dubai—surfaced in cable service magazines.[76] Although the cable colony had racialized the boundary of the network, positioning indigenous inhabitants as outside even while they moved within the station, this figuration proved dangerous to the postcolonial cable station.

Cable companies embarked on a set of strategies of interconnection to localize staff, a strategy of protection that would keep these stations from being seen as an oppositional force or a gateway (or potential interception point) to colonial power. Although at the outset of the coaxial period, the staff consisted mostly of expatriates, AT&T eventually began to hire local workers in places like Guam and Hawai'i.[77] Cable & Wireless engaged in efforts of "Nigerianization," "Bahrainisation," and so on, bringing students from over thirty countries to train at the Porthcurno Engineering College in England and establishing overseas training schools in places such as Aden, Bahrain, Barbados, Brazil, Fiji, and Hong Kong.[78] People from one part of a cable network would be invited to installations elsewhere in the system for training: for example, during COMPAC's development, invitations were sent to engineers from Malaysia and Singapore to travel to New Zealand to obtain experience in cable installation and maintenance.[79] Managers of Cable & Wireless also engaged in activities that would demonstrate their support of local people. For example, the chairman of the company gave silver cups to the first Gilbertese twins born on Fanning Island.[80]

This was a significant transition in both cable station policies and discourse. Cable & Wireless men had previously been seen as representatives of the British Empire and ambassadors to foreign cultures. In an address at the Cable & Wireless Engineering School in 1950, the chairman stated: "When you go out from here to the stations overseas you will be working with men of some 56 other races, employed on our local staffs. You will often find yourselves one of a few Englishmen in a strange land. We want you to be understanding of men of other races and to realise that they will probably form the same judgement

FIGURE 3.6. Gilbertese workers, Fanning Island, 1960s. From *Zodiac*, © Cable & Wireless Communications 2013, by kind permission of Porthcurno Telegraph Museum.

of the British nation as they form of you."[81] Through the coaxial era, particularly in the Cable & Wireless staff magazine, there was a shift to a discourse of multiculturalism that newly identified local residents as part of the network. Replacing the images of the hostile or bumbling Other were descriptions of foreign crews that identified them as part the cablemen's culture. Authors expressed pride, for example, in the fact that 96 percent of the Cable & Wireless staff in Singapore were local men.[82] Articles documented the hard work of a Fijian cable ship crew, Fijian cablemen at the Pacific Games, and the travels of a woman from Sierra Leone to England, where she was "made to feel at home."[83]

Another article focused on the Gilbertese workers of Fanning Island, telling readers that the "Gilbertese found our Western standards of accuracy puzzling at first, but can now work down to hundredths of an inch" and featuring an image of the men in front of a Cable & Wireless sign, now a part of the company (figure 3.6).[84] These texts continued to use race to codify and characterize locals, often emphasizing the virile bodies of the men and the seductive exoticism of the women. For example, the cover of the April 1961 issue of the *Zodiac* displayed a photograph of the managing director of Cable & Wireless linking arms with Maori girls in traditional dress. Though unrelated to cable business, these discourses staged and restaged the integration of the cable companies with local contexts, positioning them as part of the cable community.

The cable station building itself, the key geopolitical site for station activity, became a place for staging integration as well. A recreational area was built at the Tumon Bay station, where the workers from the American and Australian stations would play both cricket and basketball. Although the activities going on inside station buildings remained secret, in many places, especially along the Commonwealth routes, the buildings were visible. The American station on Guam, for example, had a large AT&T sign in front of the building, and flags were raised to commemorate the initial landing of the cable. At times

FIGURE 3.7. Mural inside the Takapuna, New Zealand, cable station. Courtesy of Archives New Zealand. COMPAC Commonwealth Trans Pacific Submarine Telephone Cable [Archives Reference: AAMF W3327 457] Archives New Zealand The Department of Internal Affaris Te Tari Taiwhenua.

stations were celebrated as local or national achievements. At the commemoration of the Australian Overseas Telecommunications Company station at Tumon Bay, Guam's governor stated: "The station is a most welcome addition to our island, both functional and beautiful in its thin-shell, barrel vault roof design—the first of its kind on Guam . . . this roof contains exactly the same amount of concrete as a flat roof, yet is vastly more pleasing to the eye."[85] Like the Keawa'ula station, the Tumon Bay station was designed by a local architect, used local labor and materials for its construction, and was applauded for its connections to its environment.[86] Inside the Takapuna station, a mosaic mural greeted the cablemen who entered. Copied from a painting by the New Zealand artist Mervyn Taylor, the mural visualizes the island's origin story, in which a young Maori boy pulls up the North Island, articulating a connection between cable and nation (figure 3.7). When the COMPAC cable was commemorated in Fiji, the governor spoke to an audience at the cable station about its benefits for the country, including expanded employment in the station and on the *Retriever*, as well as Fiji's historical importance in international communications.[87] At the opening of the ANZCAN cable station in Port Alberni, the executive vice president of operations for Teleglobe Canada celebrated the creation of eleven full-time jobs as a "minor achievement for [the] community."[88] The address, costs of construction, and photographs of the new station had been printed earlier in the local newspaper.[89] Disseminating knowledge about the station was not considered a security breach, as long as the walls remained protected and access regulated. Members of the public were allowed inside, but infrequently and only under specified conditions.[90] Such community integration was a broader trend across the telecommunications landscape, but these ceremonies and discourses, like the celebration of local workers in the *Zodiac*, served a specific function for the undersea cable industry: they presented the network as a local or national—rather than a colonial—infrastructure and, in doing so, helped ground international circuits in the environment, easing any potential tension between nations.

The Cold War and the tumult of decolonization formed the backdrop for the expansion of the undersea telephone cable network, against which the network's gateways were insulated. Especially for Americans, there was an increased concern with fortifying stations in reaction to the interference that could be created by nuclear bombs or hostile foreign powers. In the United States the cables, like much of the architecture of the Cold War, came to be tucked away in rural areas, secured from the threat of anchors and the potential targeting of major cities. Whereas to enter into the network in the telegraph era meant performing as a cableman and becoming part of the circuit, during the coaxial period the building itself was redefined as the gateway needing protection and was regulated not through performance, but via architectural and technological access. The cable community continued to function as a critical support system keeping these buildings secure, drawing on the earlier intimacy of the cable colony. In light of the changing postcolonial landscape, especially in the former British colonies, the composition of this community changed, and new strategies of interconnection—which presented the cable station in relation to local achievements and developments, as well as international cooperation—grounded the cable system in its new cultural environment.

Fiber-Optic: Secure Visions

By the late 1980s, when fiber-optic cables were developed, the Cold War approach of hardening had been abandoned. As with the spatial deployment underlying the cable colony and the British All-Red Line, the costs of this strategy of insulation became harder to justify since the imagined sources of interference had not materialized. As one station technician quipped, "The bombs became too big to protect against anyway."[91] Because of the investments in cable gateways, these sites retained their traction, in many cases becoming the foundation for global Internet infrastructure. Station expansions are often built atop the concrete foundations of nuclear fallout shelters. The decontamination chamber of Tanguisson is now a closet, but the station's kitchen remains empty and unstocked. What constitutes a mode of insulation in one cultural environment can make flows of power vulnerable in another. The lowest depth of one underground station is below sea level: hardening makes it susceptible to hurricanes and a rising ocean, but the area's cabled topography, including its numerous connections with other networks, holds the node firmly in place.

These layers of insulation are less often a product of national interests than they are the result of a privatized and competitive environment that values reliability.[92] Given the amount of traffic and money that traverses cables, losing a line for even a short time could eventually cost companies millions of dollars. The revenue generated per circuit is declining, and money lost when a cable is broken will never be made back. In the monopoly period, customers would have to take whatever level of service was provided, but with today's network, what level of security to provide is a commercial decision made to meet the high expectations of the international telecommunications market. Stations are still designed to be completely self-sufficient if necessary. As one operations manager explained to me as he led me around his cable station, the "whole building is N+1"—meaning that it has a completely redundant system ready to take over if the first system fails.[93] Power is supplied from both ends of the cable: either end can sustain the entire system if the other goes out. Spares of critical equipment are kept in case of an emergency. My interviewee pointed out that every cabinet in the room is reached by two different routes; if any wire was accidentally cut, nobody would lose service. The station has systems to maintain the quality of the air, the level of humidity, and the temperature, making it suitable for electronic equipment, and its walls can withstand fire for two hours. These are all strategies that seek to insulate the system's interior environments. It is not uncommon for these support systems to determine the limits of spatial expansion for the cable network. In some stations, there is room for more cable terminals and operators, but not for another power generator or battery plant.

Despite their relative self-sufficiency, stations remain dependent on local infrastructures.[94] Beyond the power generators (which have a finite period of operation), stations rely on the electrical grid to power the movement of data as well as the numerous systems that support the building's functioning. One cable station manager tells me that his station's power bill can be $25,000 a month.[95] A recent report estimates that the annual cost to maintain cables ranges from $100 to $1,000 per kilometer—which amounts to millions of dollars per year.[96] Some of this money, as in the Cold War era, is distributed to local workers and systems. One engineer described in detail to me how local knowledge can be extremely valuable.[97] He recounted a time when he and his colleagues received "local information" about how to better position the diesel generators for a fiber-optic cable project that was being established in a cyclone-prone area; in another instance, they received helpful suggestions as to how best to cross a stream. In each of these cases, local knowledge of the environment saved them both time and money.

As the era of intensely fortified superstations came to a close, two new models of the cable station emerged, which together reveal a shift away from the architecture of the building itself and toward strategic connections with new sets of circulations. The first of these is a superstation even more grandiose than the Cold War architectures. As one cable builder describes it, drawing on terminology from the American Cold War, "They are all built like they've been designed to withstand a nuclear winter."[98] I first visited one of these on Guam, which was financed during the cable boom and built in 2002 to land the Tyco Transpacific Network. The building is enormous. However, when I walk in the reception desk remains empty, occupied only by a small security camera. I write my name and the date on a sign-in sheet. To enter, I have had to gain clearance ahead of time. Policies of screening access continue from earlier eras: the U.S. government screens employees who work in U.S. cable stations (who must be American citizens), and at times project managers for cable networks from other countries (it is presumed that American citizens will be more reliable in case the government needs to close the gateway). From the entryway, I move into a large, hollow room containing rows of empty cubicles. In a tour around the building, I wander down cavernous hallways and through expansive rooms. On one side of the hallway is a giant battery plant in which everything is red; this is the red plant. Round plate batteries are stacked in many rows. On the other side of the hallway is another version of the same plant, but blue; this is the blue plant. The station stands in contrast to the clustered, layered, and remade spaces of the Cold War buildings. There are no photos on the wall, or anything to mark the fact that we are on Guam. The station has been meticulously designed, its equipment is completely redundant, and it has more than enough power and space to land a number of other cables. Like many other stations, it was overbuilt in anticipation of future growth; after its construction, the market crashed and no other networks landed here until 2009.

The Guam station and others like it have extra room that they can offer to other cable systems or rent as colocation space, in which different companies can run data centers, call centers, or services for digital content generation. This is possible for three reasons. First, there has been a shift in cable technology. Although earlier coaxial cable equipment was big and bulky, each generation of cable equipment is more efficient (yet also more complex) and smaller in size than the previous one, which frees up more and more space in the station. Second, there has also been a shift in protection away from access to the building itself: having a dedicated facility for a cable station is no longer critical. And third, there has been a shift in management. In the coaxial period, a single incumbent telecommunications company controlled each station, but privatization opened stations to a range of players, allowing new companies to colocate there. Together, the insular environment, the connectivity of the gateway, and the decreasing spatial needs of electronics have made the cable station site an ideal place for generating new forms of data traffic.

As companies come into the station to interconnect their traffic, there has been a return to the spatial organization of the telegraph station, which was a place where different groups intermingled. During one station tour, a technician points out to me the French, German, and Japanese equipment and tells me about the many people from different nations whom he talks with on a daily basis: "It is the United Nations of communications," he tells me.[99] Since the 1990s scientific cables have also terminated here, and marine scientists enter to check on computers recording subsea data. This mixing at the station often means that, like the coaxial stations, the interior is segmented. One cable operations manager tells me that he and his colleagues have tried to make their station "a bit more customer friendly" for the other groups who share the space by installing a refrigerator and kitchen, but at the same time they keep the office secure by putting in doors that require different keystrokes. He points out a personal computer behind one door and says "if you know the log-in for it, it gives you the authority to turn off all the wavelengths on the submarine cable system and shut the system down."[100] He then identifies another site to be guarded: "On the bookshelves we've got all the manuals and technical documentation for the system. If somebody wanted to get a hold of that, it would not be very good."[101] If in the Cold War era, the cable station's walls were the boundary between the network's inside and outside, today this boundary is defined in terms of actionable knowledge: keystrokes, log-in codes, and documentation manuals are the materials to be guarded. Access to the station is regulated rather than blocked, and the station becomes permeable to new kinds of bodies, a place where colocated companies can connect with multiple flows of information.

At the opposite end of the spectrum from the superstation is a modern version of the cable hut, smaller in scale and built to be just big enough for the capacity running through it. One engineer describes these stations to me as "functional equipment buildings," and from the outside they often look like windowless office buildings, blending into an industrial landscape.[102] When I visit one cable station in Hawai'i, there is no automatic gate. Instead, the station manager has to come outside to unlock it and let me in. The building is oriented along a horizontal axis. We walk by a few small offices and a bathroom, arriving at a conference room stuffed with binders and papers. Like all of the other stations, this one has standard equipment: racks of servers, power generators, and air conditioners: this tour is quick. The manager observes that even though most of the operations have been outsourced to network operations centers, the company will always need someone to conduct the environmental monitoring—scanning the horizon for ships that could anchor nearby and keeping tabs on any local development.

The extreme version of the cable hut model is made possible by a series of key technological changes. For older systems, companies had to put the cable's power-feeding technology and signal termination equipment in one place, and since they wanted to power the cable from as close to the ocean as possible, this resulted in expensive rural cable stations. In the fiber-optic era, these two technologies can be separated. The power-feeding equipment can be located in a small hut by the shore, with the rest of the equipment in an urban hub. This transition effectively means that the cable station can now be just a power station—a change that telecommunications managers see as a dismantling of the "big AT&T model."[103] The diagrams for undersea networks now often extend beyond the station to the point of presence—the place where the signal can be accessed, which is often in an urban area. Moreover, the development of the undersea branching unit, a technology that can connect three cables on the seafloor (meaning that if any one cable station was taken out, the other two would still function), also made the cable station less critical as a hub for signal traffic. These new models—the superstation and the cable hut—disperse part of the activity to other locations.

As in earlier periods, today interference is envisioned as an interception of the signal at the cable station, but strategies of insulation now focus on protecting the station from tactical terrorist attacks on (or close to) the ground—a process that moves regulation to the environment surrounding the station, or the buffer zone. Thinking about the security of the station, one cable entrepreneur narrates his thought process: "Can it take a light plane crash? It's got a really heavy-duty double-skinned roof. Can it take an eighty-kilometer-an-hour twenty-ton truck? Yes, it can because of the way it's been constructed. What if someone decided to take you out? Could they?"[104] Immediately after 9/11, the U.S. military surrounded several cable stations on American soil because they were considered a possible target. The intensity of surveillance increased, as did the number of cameras monitoring the stations. At some stations in the United States, the Department of Homeland Security conducts site assessments to suggest increased security measures. On the list of changes that the department would like the owners of one cable station to make, for example, is clearing out the trees between the station and the cable's route to the ocean, to enable better visual surveillance of the route. At another station, the department recommended moving the road back from the building, making it more difficult to access from the street. These strategies are more effective in manipulating a station's field of vision than in actually keeping people out: instead of constructing the station as a point of national pride, it must now be insulated from public perception. Today's cable stations often remain unmarked and without signs; some do not list their addresses. At times, stations are indicated in policy documents only by coordinates. Of the eight vulnerabilities to stations listed in the ROGUCCI *Study Final Report*, an analysis of the fiber-optic undersea cable network, two are explicitly linked to the stations' visibility: they are identifiable and subject to external surveillance.[105] Together, strategies of insulation simultaneously increase the knowledge that cable companies have about stations' buffer zones and decrease the knowledge that outsiders can obtain about cable infrastructures.

As in the Cold War, today the desire to create a buffer zone has to be weighed against the affordances of the existing topography. About fifteen minutes by bus from the center of a large city, one major cable station is located on a busy street lined with cafes, restaurants, and upscale retail stores. On one side of the station is a church and a school. Children play in a children's center across the street. This central location and proximity to urban life is abnormal for a contemporary cable station. When I question one of managers about the building's central location, he tells me that it is not an optimal site but that moving would be too expensive. "I guess there's less paranoia than there is in America," he suggests. "If we started to have direct strikes, then the whole ideology of how you protect your facilities would change."[106] Since he and his coworkers cannot move the station, they attempt to blend into the environment and distract attention from the building's function. At a different station, a cable entrepreneur echoes this sentiment, saying that the goal there is to "make the building as nondescript as possible."[107] In contrast, the cable station manager tells me that a second facility they run is secure because it is located in a fairly isolated area, with high fences, barbed wire, and an automatic gate with video cameras operated 24/7 by staff members who will only open it if they recognize the visitor. There are about sixty actively monitored video cameras, and the police regularly visit to make sure that nothing is out of order. At night barriers block off the roadway, giving the station "multiple layers of protection."[108] Those who take this approach run the risk of alerting the public to the importance of the facility: with all of the high walls, guards, and electric fences, the manager comments, "You know it must be a really good target."[109] In the United States, he says it is very easy to spot critical infrastructure: the stations look like "special fortified bunkers."[110] He speculates that if their company suddenly bricked up the windows of the urban station, reinforced the facility, and posted "Restricted Area" signs, it would actually attract more attention. In these two opposing strategies—making the station invisible to the public and making the surrounding environment visible to the station—the terrain of security, the buffer zone, is a visual one, and it is negotiated in the zone around the station.

Although some companies continue to use the model of "security through obscurity," others have attempted to use cable visibilities to develop lines of communication with potential customers.[111] Pipe International developed a blog for its PPC-1 cable, not only following the cable laying in a narrative of connection, but also revealing the endpoint equipment in the cable station and the regulatory environments that the cable must pass through. The company even used video to address questions about the system's features and used its website for public dialogue about cable technology. This was important for its market exposure. When staff members designed the blog, they were conscious of manipulating the visibility of the station. A company representative remembers that they knew they could not be completely open, because they didn't want "an army of bystanders" outside the door: "You tell them just enough so they can feel part of it."[112] When the company put photos on the blog, these were often close-ups with a low level of detail, the text was illegible on screenshots, and images were compressed with low resolution, so that people couldn't use them for data-gathering exercises.[113] Since such imagery could compromise security, the company made sure not to post information that would give away the station's precise location.

Another set of transitions helped move activity outside the cable station: as the manual routines from the coaxial era were automated, remote signal-operation and control systems were advanced; and as cable systems grew more complex and interconnected, there emerged an increasing need for network management. Much of the labor of a cable station was relocated to a network operations center or a network management center.[114] At the same time, massive corporate downsizings, which occurred across the telecommunications industry, began to replace workers with computers. As a result, cable stations—both superstations and cable huts—today might be staffed part time or not at all, with workers called in only when they are needed. Some technicians argue that unmanned cable stations are one of the most critical vulnerabilities of the system; they tell me that having experienced technicians on site to respond immediately in the event of a failure could save their companies millions of dollars. Others have embraced the change. They tell me instead that many failures are minor and don't require an advanced technician to identify them.[115] Nonetheless, these supervisory systems mean that the labor of cable maintenance, once concentrated in the cable station, now takes place in a variety of locales.

A more detailed description of cable labor illustrates the continued importance and strategic value of cablemen's knowledge to the functioning of information networks, despite the reductions in the workforce. Today, cable labor consists of actively monitoring and preemptively fixing errors and alarms before they turn into problems. In a network operations center, the interior is dominated by computer screens with information about the network's operations and specifications that is continually updated. The daily work of men (and men still make up the vast majority of the cable workforce) in these centers entails addressing a series of alarms. An alarm might be anything from an indication that the cable has stopped functioning to a small system update. In one network operations center, a technician tells me that he and his colleagues will get approximately 120–150 alarms a week, but the number varies widely depending on the scale of the network.[116] The vast majority of these are warning alarms, which tell the cable workers that they are getting close to a threshold for action, that there is a problem with a backup system, or that there is potential interference that has not yet compromised the flow of information. It is the task of the men at the operations center to read these sets of alarms and determine what needs to be fixed, all of which might take place without a drop in signal transmission. Nonetheless, the system is always already in a state of alarm.

The variety of alarms speaks to the fundamental materiality of the cable system. Alarms can be generated in the cable station by anyone who enters: technicians, customers, and even cleaning crews. With the number of air conditioners blowing dust around, cable stations must employ specialized cleaning crews. Stations typically have an increased number of alarms during the cleaning process, and during holiday periods like Christmas the number drops dramatically. One cable operations manager tells me that "there are times when people are staying away from the network, so when you haven't got people touching stuff it tends not to break."[117] The inside of his station testifies to this precariousness. For example, the primary fibers running in from the sea are labeled with bright tape reading "Danger Optical Fiber." He tells me that this is to make sure that, in case "someone comes in and doesn't know what they're doing, they don't touch it."[118] Human bodies, whose circulation is necessary for the operation of the station, inevitably bump, jostle, and set equipment into an alarm state. During Super Bowl weekend, one company planned not to have any activity in its station just to ensure that nothing went wrong. Disruptions are far from regular or predictable, the operations manager tells me: "You'll have months where nothing happens, and then you'll have a week when you're called in every night. That seems to be the way that it goes."[119] Another engineer assures me that the major problems always occur at night or on the weekend.[120] Like cablemen throughout history, in their rhythms and experience of time, these workers live in a global temporality, often laboring in accordance with fluctuations in bodily traffic at distant locations. There is a continued need for interconnection at the station (for maintenance, repair, and cleaning), yet these interconnections simultaneously compromise the flow of information.

Further demonstrating the materiality of the system is the fact that particular pieces of equipment, even though they are supposed to be identical, each have their own fickle behaviors. Each piece of undersea cable equipment has been manufactured using different batches of raw materials and assembled at different times. Two circuit packs might be technically identical but might function differently over the course of their lifetime, in part because different computers contain materially different components. The glass or the solder wire might have been of a different quality and come from a different origin. One manager tells me that this can result in "batch faults," which occur in a series of equipment manufactured at the same time. He uses an analogy to explain the process: "It's a bit like making a fruit cake. I can make a fruit cake on Monday and I can make one on Wednesday, but they can be different even if I followed the same recipe. In the one on Monday I might have used 198 grams of sugar and the one on Wednesday I might have had 205 grams of sugar. Very, very minor differences could have an unknown impact sometime in the future."[121] The workers keep detailed records on the pieces of equipment so they know what each part's history is—"what it's been through."[122]

Some parts will develop bugs, and others won't. One manager tells me that their station just hadn't gotten the right pieces of transmission equipment, and once one piece had started to have bugs, it required repeated maintenance for most of its life. Another describes a problem he and his colleagues had with a piece of equipment that was displaying an alarm state, although the alarm was not detected at the network operations center. As a result, the men in the operations center could not determine where the bug was: in the piece of equipment in the station or their computers at the center. Even the smallest discordances need to be addressed in such a critical system. The engineer and his colleagues decided to send the equipment out (at great cost) to have its code rewritten. He tells me: "Things can go wrong and they will go wrong, but if we can lower the risk by investing more money we will do that."[123] In developing new networks, planners depend on the grounded knowledge of operators who are familiar with particular pieces of equipment. One engineer tells me that his company always involved the "operational people" in the early days of planning its systems: "They were the guys who were ultimately going to give it to you. They can contribute from their own experience as to what you shouldn't be having new or you don't want this time."[124] He emphasizes the importance of this dialogue, without which it would be difficult to get the operators to establish continuity between systems.

Although some alarms can be dealt with remotely, many of them require visits to the actual pieces of equipment at the cable station. The workers' embodied knowledge still plays a key role in the operation of the network. As much of this labor entails responding to disruptions, technicians are often on call 24/7. Some stations seem to be in a state of potential alarm at all times. Throughout some U.S. stations, for example, electronic signs hang from the ceiling with red lettering flashing up on them. These are meant to warn the employees at a moment's notice if anything is going wrong, as are loud buzzers that were instituted after 9/11. When they are not responding to alarms, technicians check on sets of scheduled events, conduct verifications, and write reports. One technician lets me follow him to a cable station on a routine follow-up to a warning alarm. He explains that these systems, just like any computer, are in need of constant updating. There is not a one-to-one correspondence between each alarm and an actual problem with the system. Rather, an alarm is a symptom that something is wrong, he tells me—an indication of a failed connection. It does not indicate what caused the failure. For example, a full cable break might generate many alarms. In turn, multiple problems might contribute to a single alarm. There is a significant amount of interpretive work required of the cable technicians in the network operations center, as they read through alarms to deduce the origin of the problem. Pointing to one rack, which has a light on, the technician says, "See, for example, that machine is in a state of alarm." He plugs in his computer to see what he can determine, but the problem remains unclear. He then turns to a rack from which several cords extend, plugging into another machine. He looks at several loose cords. "I think that this one here," he says, picking up a cord, "is supposed to be in here"—he points to a jack—"but I'm not sure."[125] He's not ready to risk it. This alarm is only for a backup machine, so it can wait. We leave the station, still not quite sure what the problem is, and head back to the network operations center.

Cable technicians' knowledge of the system as a whole, of the particular pieces of equipment, and of the history of a cable's operation is incredibly valuable. The circulation of this knowledge keeps the network itself in operation. No one person has an understanding of the entire system. Even in this station, new servers and stacks have been added, and the technician I interviewed was not familiar with the history of every single one. A large part of the contract for a new cable system has to do with training and documentation. This part has to be well managed, given the few people employed and the extensive knowledge that is needed to effectively maintain and operate the system; continuity has to be maintained throughout its life. Since it is impossible for anyone to see the network in its entirety, engineers and technicians depend heavily on each other for solving problems: they must know who to call for what information. Moreover, relatively few skilled cable workers travel around the world to coordinate these systems.[126] The connectivity of the cable community supports this interpretive work. As in the early days, the smooth operation of global communications often rests on the abilities of a few men to act quickly and minimize cable disruptions.

The level of secrecy of the job, the specialized nature of cable work, and the relatively few number of cable systems has produced a fairly small and insular group of cable industry workers. The downsizing of the 1990s and 2000s led to a further contraction and concentration of knowledge within this community. At most of the stations I visited in the Pacific, managers and technicians all know each other. One engineer described the situation this way to me: "When a friend goes to work on another project, I know what project he's working on even though he doesn't ring me up. I know he's involved in a project in Singapore because the guy in Singapore told me. Probably shouldn't have told me, but I know. Everybody seems to know everything."[127] The importance of knowledge and peer respect in the industry is absolutely critical. The engineer continued: "A number of times someone has said to me, 'I'm going to send you a file to explain the situation. You're not supposed to have the file . . . please don't say you got it from me. But you need to read this before you make a decision.'"[128]

Despite the station's decreasing geopolitical significance in the fiber-optic era and the dispersion away from its built architecture, a dense social network continues to support the technical network, a cable community sustained via a common set of experiences and geographies, a shared history, and continued interactions at the cable station and other network sites. In my interviews I am often surprised by the affective relationship that cablemen have to these jobs. Many speak fondly of their time working in the industry. In more than one interview, when a cableman has left the room to gather more materials, from plates commemorating landings to pieces of undersea cable stored in closets and basements, his wife has told me the extent to which he became involved in his work. This sense of shared history and purpose is discernible even in cable stations today. When I visit the station at Tanguisson Beach, on Guam, I see on the wall that there are yearbook-style photos of all the AT&T stations in the Pacific: Makaha, Keawaʻula, Bandon, San Luis Obispo, and Manchester. Each page has an image of the station and photos of the people who worked there, with their name and position listed underneath, grouped as a school class would be. Photos of the cable ships and of AT&T cable landings and maps of Guam decorate the walls. At other cable stations, there are platforms that display segments of the coaxial and fiber-optic cables that have run through the station. When I visit a cable station in Fiji, I find a sense of camaraderie and hospitality among the cable workers. Several men are hanging out around the station: it has been a training ground for local communications workers and a site for teleconferencing, and it has a large courtyard with a monument commemorating the development of the site as a satellite earth station and the landings of the ANZCAN cable in 1984 and the Southern Cross Cable in 2001. Here, the station's relatively open spaces facilitate the connection of cable communities with other local publics and make the cable visible as an object of pride.

These connections within the cable industry are partly due to the strong sense of culture and family that has been present since the telegraph era, even if the cable station is no longer the primary site at which this identity is formed. Commenting on Cable & Wireless, the company's chairman writes: "The company is, in fact, a huge continuing international family."[129] A cableman reflecting on his time at the Port Alberni cable station suggests that the cable system "was like an extended family. Wherever you went in the world, if you visited a cable station, you were bound to meet someone you knew or had heard about. This could be a very reassuring connection in time of need and was always a wonderful opportunity to do a bit of reminiscing."[130] At a recent telecommunications conference, Fiona Beck, the CEO of Southern Cross Cable Network, was asked: "What is significant for you about the industry?" She responded: "For me, the industry is like a family."[131] Dean Veverka, chairman of the International Cable Protection Committee, reiterated this description in an interview: "In the industry, we compete, but we also share resources. It's competitive, but there's a family nature."[132] Although these descriptions of the industry as a family naturalize its historically monopolistic structure, for many cable workers they serve as an apt metaphor. Ultimately, cable companies must coordinate with one another to reroute service in case of a break, to connect with each other's networks, and to access a limited pool of maintenance resources: if one company's cable goes down, the cable ship will not be available for anyone else's repairs. The technicians, operators, engineers, and managers respect each other's work and view the construction and operation of global cable networks on the whole as a collaborative activity, work that they take pride in.

In an interview, Beck and Veverka described to me the strategies they used at Southern Cross to cultivate a sense of community and make the company a "tight-knit team."[133] In part, they attributed the cohesion to the magnitude of the project: the company spent over a billion dollars to serve millions of people, and "everyone had a common goal, there was a real sense of shared vision."[134] From the very beginning, they spent time bringing the team together in various locales around the world, making sure to include the local partners and landing parties. These face-to-face contacts and knowing about another person's family, interests, and personal life is important from an operational perspective, Veverka said, because when something goes wrong, "you can pick up the phone and you know the person at the other end, you've spent time with them, face to face, it's quite unique in a sense. You trust them."[135] He argues that in the case of cable failures or breaks, without this trust workers would be calling too often or with the wrong questions. Southern Cross celebrated anniversaries and held retreats, trying to make them fun, and whenever possible invited workers' partners and families along, a practice that helped employees become a more cohesive unit. For Southern Cross's tenth anniversary, the company staged a version of the Amazing Race for the cable workers, challenging them to find their way across the New Zealand part of the cable route. These activities might seem peripheral to cable work, but they helped build a social infrastructure, oriented around the network's geography, that is absolutely critical to the cable's functioning.

When I ask cable operators about the largest vulnerability of the current system, many express concerns about downsizing and retirements: they fear that carefully guarded industry knowledge will be lost and that there will be nobody to take their place who will adhere to the same standards of reliability. Recruiting the next generation of workers is difficult, as there is no direct path to the cable industry and it remains largely invisible to the public. As one engineer describes the situation, "Nobody goes to school and says I want to be in the undersea cable business."[136] Although turnover in the technology industries has been increasing, many cable workers have been in their industry for decades. One woman estimates that 90 percent of her co-workers have been in it for over a decade, and "even if they change business cards and move from Cable & Wireless to BT [British Telecom] to Alcatel, everyone still knows everyone."[137] After cable workers retire, they often remain in the industry as consultants, and they still have access to information about cable networks, installations, and operations. In a privatized system that values outsourced labor, this knowledge keeps them a part of the cable network. The embodied knowledge of the cablemen, built from their experience, is one of the greatest assets in the system. In helping maintain the continuity of global flows of information, it constitutes part of the cable's insulation.

In the fiber-optic period, the boundary of the network is therefore negotiated and policed on the terrain of actionable knowledge. If in the early telegraph era to be part of the network was to perform as a cableman, and in the Cold War era to become part of the network was to gain access to the building, in the privatized cable system to know the network's operations is to be part of the network. Strategies of insulation—from security cameras, blending in, and regulating the buffer zone to monitoring cable station keystrokes and system alarms—keep the network secure by amassing information about its activities and letting very little of this information out. It is in this information surplus that security is created. As always, interconnection with local power grids and local publics is needed, and the transoceanic cable network has been produced by and is dependent on a small group of people who are intimately connected to the system and who have a shared sense of purpose in upholding global communications. The construction and upkeep of these systems is profoundly material. Without the work of daily troubleshooting, our international networks would eventually fail. It is the narrow channeling of knowledge through specific, located bodies that has made possible and continues to support the fluidity of global information flows. The cable network's greatest vulnerability today, in the wake of the monopoly system, remains a lack of channels for circulating cable knowledge toward the production of a reliable next generation of undersea networks.

Erasing Cable History

The Passing of a Cable Station should not go unrecorded and unsung, yet so we allowed La Perouse to vanish out of our official life.—C. Holroyd-Doveton, untitled article

Given the historical importance of the cable station to contemporary information networks—every era of cable infrastructure has been built in relation to the previous one—I track down the original telegraph cable stations of the Pacific. I end up at the U.S. Naval Base in Guam, standing at the edge of the ruins of the original Sumay telegraph cable station, next to the base's community liaison. The air is humid, and clouds stack up on the horizon, one on top of another. "The last time I was here," she tells me, "it was all overgrown. This path wasn't here and we had to stomp through the jungle."[138] She has kindly arranged for me to visit the station, inaccessible without a pass and an escort. The cable at Sumay operated for from 1902 to 1951—almost half a century—and was part of a secure line of communication connecting east and west, linking the Philippines, Guam, and Japan to the mainland United States via Hawai'i. Given this station's role in keeping Guam central in transpacific networking, and because I had just visited Guam's three fiber-optic stations, I had expected that the ruins would be better marked. Many other sites bombed during World War II are memorialized across Guam, but all that exists here is a single sign testifying to the station's existence. We move along a recently cut path through a maze of concrete structures overtaken by the jungle. It is only now that a team is beginning to excavate ruins in the area from over fifty years ago, though it is not clear to my guide who the team's members are or what their specific task is. Using archival images from the years prior to World War II, we can only guess as to which of the buildings remain (figure 3.8).

The stations along the American transpacific telegraph cable routes remain unexcavated, unknown, and insignificant in our historical landscape. Although the ruins at Guam are commemorated, they are located on the military base and are not accessible to the public. At Ocean Beach in San Francisco, though numerous pictures were taken to document the original cable landing, there are no remnants of these infrastructures today or indications of their historical role in global telecommunications. The original cable hut has been demolished, and an apartment building has been put up in its place. Like Ocean Beach, Hawai'i's Sans Souci Beach is a major tourist beach in downtown Waikiki, yet the only remnant of the cable is a brief mention at the bottom of the menu in a local restaurant. Much of Manila's coastal shoreline has been built out into the bay. The telegraph landing point here is probably submerged beneath roads and shopping malls. The ruins at Midway Island cannot be accessed since there are

FIGURE 3.8. Ruins of Sumay cable station, Guam.

no regular commercial flights to the atoll. Only military personnel, research teams, and ecological expeditions that charter flights may visit. The remnants of transpacific communication histories live on solely in the historical images of cable landings and commemoration ceremonies, scattered few and far between and often buried in remote archives.

The spatial development of the Pacific Cable Board and Eastern Extension routes that linked the British colonies contrasts with that of the American routes. Many cable station buildings along this line have been repurposed for community and scientific uses. The station at Bamfield has been transformed into a Marine Sciences Research Center. Scientists seeking new species near Wellington, New Zealand, briefly used the cable house at Titahi Bay.[139] For a period of time, the University of Hawai'i's Oceanographic Department leased the cable facility at Fanning Island, and the cable stations were later turned into schoolhouses.[140] The original telegraph building in Fiji has been renovated and restored to house Fiji International Telecommunications Limited (FINTEL). In Southport, Australia, a park, a street, and an apartment building are named after the cable. In the 1980s, when the cable buildings were under threat of demolition, residents spearheaded a campaign to save them; they were renovated, restored, and moved to form the Music Department of the nearby Southport

School. The cable hut still stands and is marked in the Queensland Heritage Register as a significant site of cultural heritage. The La Perouse station, outside of Sydney, was briefly used as quarters for nurses at the Coast Hospital and later became a women's refuge operated by the Salvation Army.[141] This station is now a historical museum, as is one of the cable buildings in Darwin, Australia. The station at Cottesloe, Perth, in Western Australia, was subsequently used as a children's home. In Northland, New Zealand, the original telegraph house has been moved and was reused as the home of a local cinema called the Swamp Palace. Along the original transpacific cable route, it is not uncommon to find images of a cable landing on the walls of nearby buildings and artifacts from the telegraph era open to the public. In many places, including Southport and Darwin (Australia), Nelson and Northland (New Zealand), and Bamfield (Canada) there are historical monuments and nearby exhibitions to commemorate the cable's landing and role in the community. History is present in these sites, made visible for those who live there as well as those passing through.[142]

Most of the sites are no longer anywhere near a current cable landing. Canada's traffic is now routed through U.S. cables. Fanning Island has no cables. Much of Australia's Internet traffic is funneled through central or northern Sydney rather than Southport, La Perouse, or Darwin. New Zealand's Internet traffic leaves the country through Takapuna, north of Auckland, rather than at two original cable landing points (both named Cable Bay) on the tip of North Island and the tip of South Island. The visibility of cable history along the British colonial route is accompanied by a shift away from these locations as significant telecommunications hubs, while the movement to invisibility and secrecy as a strategy of insulation for American cable stations is accompanied by a lack of institutional interest in the maintenance of early cable sites. Our forgetting of cable stations, however unintentional, might serve as a strategy of insulation for contemporary networks. If we do not think about the historical telegraph or coaxial cables in an area, it is unlikely that we will think of the contemporary cables and will not see them as a potential target for disruption.

At stake in the forgetting of cable history and the decreasing visibility of communication infrastructure is our understanding not only of the significance of cables, but also of the significance of their ongoing cost and the embodied difficulties of operating international networks. Once networks are in place, they do not just operate smoothly. Barty-King observes that "the easier part of the exercise was laying a cable and setting up the service. . . . More difficult was what they laconically called 'maintaining the cable,' that is keeping it open for public traffic without interruption."[143] It is at the gateways of our networks that such investments in maintenance are made, cable labor is stabilized, and the boundary between what is inside and what is outside the network is defined. In each period this occurred in relation to the contemporary environmental context. As a result, stations were imbricated with different historical models of security, moving from the colonial era, in which protecting the network entailed regulating the cablemen's bodies; to the Cold War, in which security was displaced to the built architecture of the station; and then to the fiber-optic era, in which security entails monitoring, regulating, and restricting knowledge about cable infrastructure. In each of these eras, the cable station has continued to be the inhabited architecture of our global networks, the place where cosmopolitan, global workers form their own communities, connect with local cultures, and inflect global telecommunications operations. Both strategies of insulation and interconnection made it easier to lay new cables to existing stations, giving the Internet traction in the historical gateways of cable communications.

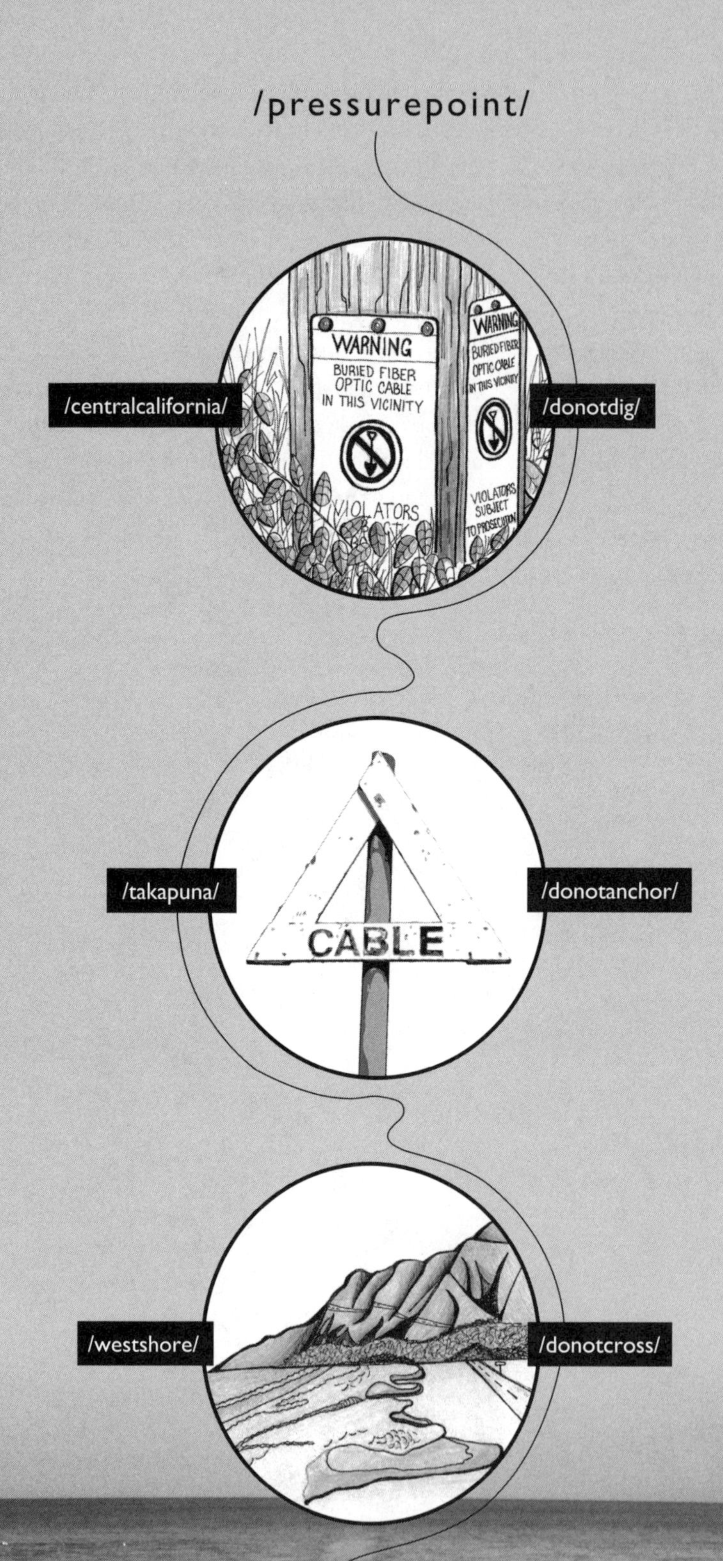

/pressurepoint/
/centralcalifornia/
WARNING
BURIED FIBER
OPTIC CABLE
IN THIS VICINITY
VIOLATORS
WARNING
BURIED FIBER
OPTIC CABLE
IN THIS VICINITY
VIOLATORS
SUBJECT
TO PROSECUTION
/donotdig/
/takapuna/
CABLE
/donotanchor/
/westshore/
/donotcross/

FOUR

pressure point

Turbulent Ecologies of the Cable Landing

Moving out from the cable station—where signals are monitored, powered, and processed—toward the deep sea, undersea cables must first traverse roads, beaches, and coastal waters inhabited by aquatic creatures and terrestrial publics. This zone is the cable landing point, described in the industry as a "transition environment."[1] Like the cable station, landing points are often displaced from urban areas and are few in number due to the cost of establishing them, connecting them with domestic infrastructure, and setting up strategies of insulation. Fewer than ten cable landing points on the west coast of the United States collectively make up the gateway for almost all international data traffic to Asia. The spatial organization of these landings diverges from that of the cable station in two key respects. First, although the cable station has been conceptualized as a potential target throughout its history, the cable landing point has remained at the periphery of security concerns. Cold War cable stations were housed in nuclear fallout shelters, but the cables extending from them were encased only in slim metal conduits. Cameras monitor the cable station; in contrast, there is rarely any electronic visual surveillance of the cable landing point. Second, although cable stations are private spaces where telecommunications companies restrict access, the coastal areas of the landing point are public spaces where the cable regularly intersects the movements of diverse actors. Whereas it took years for me to gain access to some cable stations, I did not need permission to walk over the manholes at cable landings or stand on the conduits through which networks run. The most significant threat that cables currently face derives from their routing through public space. At the cable landing, cables encounter friction from local actors invested in their own modes of spatial practice and organization. Here, as Susan Leigh Star observes, "one person's infrastructure is another person's . . . difficulty."[2] The majority of actual cable outages occur on either shallow coastal segments or on terrestrial routes. In this chapter, I describe how the cable landing is a pressure point in the system: a zone where circumscribed local practices can have a disproportionate influence on the global development of media infrastructure.

To do so, I weave together three nodal narratives that highlight the negotiations of cable companies with turbulent ecologies at the transoceanic cable landing point, surveying three dominant forms of interference for cable systems. I begin with the conflicts between telecommunications companies and environmentalists in California who are interested in keeping the coast a "natural" space and preserving its "aesthetic resources"—a source of turbulence that makes the state one of the most difficult places to land cables in the Pacific.[3] However, the most significant and long-standing disruptions to cable systems have been caused by boaters and fishermen who drop anchors or trawling gear on the cable. Comparing coastal conflicts in the early twentieth and the twenty-first centuries, I then describe the strategies of insulation that cable companies have developed to protect their property against such disruptions, although fishermen retain significant power. Conflicts have also unfolded with the coastal communities under which cables are routed. The third landing I discuss is on Hawai'i's west shore, where communities have made Farrington Highway a particularly difficult site for cable companies to cross. Although strategies of interconnection with existing circulations have been key to stabilizing the cable station, at the cable landing point local activities have more often been conceptualized as threats. In each of the cases above, I argue that companies had to mobilize strategies of insulation in order to transform turbulent ecologies into frictionless surfaces, though there have been intermittent and rare attempts to interconnect with these publics. Due to the difficulty in expanding strategies of insulation, cable companies often continue to route

FIGURE 4.1. Scuba divers near a Guam cable landing.

networks through particular sites. As a result, the cable landing point, like the cable station, is a topography that pulls in further telecommunications investment and brings unseen advantages to the surrounding area.[4]

Strategies of insulation at the cable station involve extensive manipulations of architecture and social practice, but at the cable landing point they involve specific negotiations of the cable's visibility to local publics. These visual surfacings, like conflicts over cable routes, reflect local spatial practices: scuba divers regularly use cable conduits to find their way out to a Guam reef; in California, cables are indicated by warning signs that blend into the environment; and in Hawai'i, some routes are not marked at all (figure 4.1). I close with a consideration of a cable landing in the Philippines, where the route remained invisible to me, to mark some of the limits to visualizing cable systems. This chapter, which narrates conflicts around the cable landing, reveals yet another zone of the network where operators interface with local publics and are affected by the surrounding environments, yet where a different set of actors—including unlikely players such as fishermen and conservationists—have influenced the dispersion of Internet infrastructure. These pressure points are sites where seemingly insignificant microcirculations, tiny eddies in a global system of currents, have far-reaching impacts across the ocean, and the friction-free nature of global communication is contested.

Do Not Dig: Preserving California's Aesthetic Resources

California has been critical to cable communications since the early twentieth century, when the first American transpacific cable was routed into San Francisco. Fifteen international cables have since landed on its coasts (map 4.1). Although the state has been a hub in communication technology, it has simultaneously played an important role in the growth of the environmental movement, pioneered breakthroughs in environmental policy, and facilitated the preservation of wilderness areas. Daniel Press notes the apparent contradiction in the state's attitude toward development: "At the vanguard of paradoxical trends, California is one of the fastest growing states in the country, but is the most ambivalent about change."[5] As the accelerated extension of fiber-optic cables in the late 1990s made California a gateway for information traffic

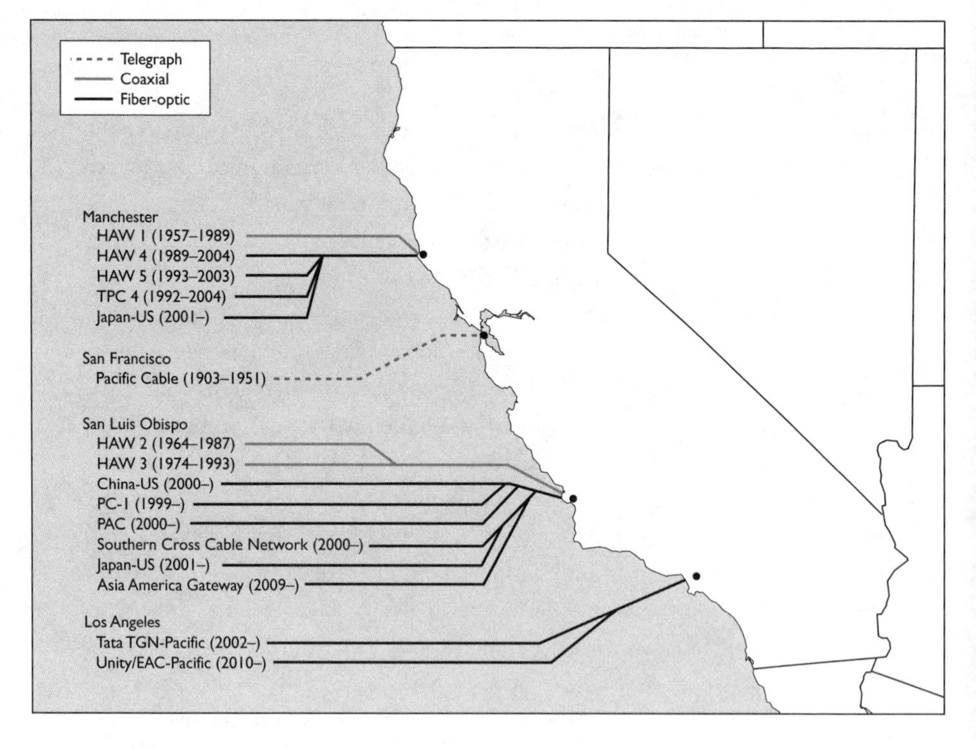

MAP 4.1. California's transpacific cable landings.

FIGURE 4.2. Manchester, California, cable station.

to Asia, the state became one of the most turbulent sites for cable laying in the Pacific. Cables had to be routed through public space, and in the process they encountered friction from environmental organizations invested in preserving the balance of existing natural ecologies.

The friction generated at California's landing points is exacerbated because all but one of them are located in rural areas, often passing through state parks—in part due to the Cold War–era infrastructural decentralization discussed in chapter 1. To get to California's oldest still operational cable landing point, established in Manchester in 1957, I drive over a hundred miles north from San Francisco on narrow wooded roads and along the Mendocino coast. The cable station is positioned on a bluff overlooking the Pacific. Across the road from the station, families picnic at a campsite. Cows graze in a nearby pasture. Between the station and the beach, a sign restricts access to a "sensitive habitat area" (figure 4.2). After Manchester, California's next landing point was set up in 1964 at Morro Bay, midway between the commercial hubs of San Francisco and Los Angeles. From its beach landing there, the cable is routed through Montaña de Oro State Park to a station nestled in the hills west of San Luis Obispo. These spaces are some of the most critical gateways to U.S. telecommunications, but they are publicly understood and regulated as natural landscapes.

At first glance these environments, and the signs indicating environmental preservation, appear to secure both wildlife and cables, protecting infrastructure from perception and therefore potential tampering. Extending cables through these ecologies, however, requires an incredible amount of effort by cable companies: acquiring environmental permits and adhering to their guidelines constitute a significant source of friction. Companies must develop Environmental Impact Reports (EIRS) and submit permit applications to numerous agencies. For example, the Southern Cross Cable Network had to secure the following prior to landing its cable:

- Coastal Development Permit and Federal Consistency Certification (California Coastal Commission)
- Lease of submerged state lands (California State Lands Commission)
- Clean Water Act Section 404 and Rivers and Harbors Act Section 10 authorizations (U.S. Army Corps of Engineers)
- Lease of park lands (California Department of Parks and Recreation)
- Coastal Development Permit (San Luis Obispo County)
- Water Quality Certification and Storm Water Certification (Regional Water Quality Control Board)
- Streambed Alteration Agreements (California Department of Fish and Game)
- Essential Fish Habitat authorization (National Marine Fisheries Service)
- Biological Assessment concurrence (National Marine Fisheries Service and U.S. Fish and Wildlife Service)[6]

Many of the government agencies involved, in particular the California State Lands Commission and the California Coastal Commission, are invested in the preservation of coastal lands and maintaining the balance of existing natural ecologies. They provide a forum where various stakeholders, including environmentalists and fishermen, can voice concerns about development, and they publicly mediate among diverse interests to minimize the environmental damage of any given project.

The permitting process, which is not coordinated across these agencies, is a lengthy, expensive, and often confusing undertaking for cable companies—a form of friction that slows development. Applicants must report on the projected effects of the proposed system as well as the potential impacts of alternative modes of installation, comparing both to a scenario in which the cable is not landed at all. The creation of EIRS involves addressing all concerns raised by the agencies and local advocates, ranging from the project's potential interference with the growth of native plant species to the emissions from the ships used to lay the cable. On the whole, the direct environmental impacts of cable installation and maintenance, even plowing a cable into the seabed, are small.[7] Although scientific knowledge has been brought in to substantiate the claims of both sides, the studies that EIRs are based on often lack definitive evidence in either direction, due to the limited research on cable systems' actual impacts. The issues addressed tend to reflect cultural interests and fears more than actual disruptions. For example, a 1950s study of whale entanglement in cables—which was based on examples from an earlier period of telegraph cable laying, during which cables were sometimes draped between undersea hills or left coiled on the seafloor—has been used to inform decisions made about coaxial and fiber-optic development.[8] The only known conflict between fiber-optic cables and large marine animals dates back to the mid-1980s, when an experimental cable laid off the African coast was severed by bites from whitetip and crocodile sharks (many of which occurred below 1,400 meters).[9] After this disruption the industry quickly adapted its networks by adding an extra insulating layer that blocked the shark-attracting frequency. Nonetheless, due to the California Coastal Commission's guidelines and the continuing public concern about dangers to such charismatic species, cable companies abide by a Marine Wildlife Contingency Plan that requires an official marine mammal spotter to be present on boats during cable laying. This serves as a strategy of insulation, which protects the companies both from whale entanglements and from the protests of environmentalists.

In many cases, the enforcement of environmental protection laws is designed to preserve California's lands in a "natural" state. This is strikingly brought into relief by a section of an EIR for the Asia-America Gateway cable titled "Aesthetics/Visual Resources." The report cites the California Coastal Act at the outset: "The scenic and visual qualities of coastal areas shall be considered and protected as a resource of public importance. Permitted development shall be sited and designed to protect views to and along the ocean and scenic coastal areas, and, where feasible, to restore and enhance visual quality in visually degraded areas. New development in highly scenic areas . . . shall be subordinate to the character of its setting."[10] The report goes on to assess the visual markings of the cable and analyze how they are "subordinate to the character of [their] setting." This analysis includes pictures taken specifically to emphasize the consistency of route markings with the existing visual landscape (figure 4.3). It considers factors such as the potential light and glare impacts of cable signs, the visual disturbance of construction, and the trimming and removal of vegetation (the analysis recommends that AT&T retain a

FIGURE 4.3. Photograph used to document aesthetic resources to be preserved, San Luis Obispo, California. From California State Lands Commission, "AT&T Asia America Gateway Fiber Optic Cable Project Final Environmental Impact Report SCH No. 2007111029."

FIGURE 4.4. Landscaping along a cable route in Guam.

certified arborist "to perform any necessary trimming of oak tree limbs overhanging equipment access routes").[11] In other locations, manholes are located underneath the grass, marked only with sensors that enable companies to detect them. By carefully embedding and disguising infrastructure in the natural environment, cable layers manipulate the route to accord with the local spatial politics of environmentalists, environmental organizations, and residents worried about lowered property values. Such strategies are location-specific and culturally inflected. For example, on Guam a significant amount of resources are allocated to route maintenance each year, including landscaping around bright yellow route markings to keep them clear of the quickly growing grass—a process that distinguishes cables from the environment to keep local development projects from unearthing them (figure 4.4). Although California's cables are disguised in the landscape, Guam's cables are insulated by visibility.

In some places environmentalists have succeeded in stopping cable development altogether. At the height of the cable boom in the late 1990s, the Southern Cross Cable Network (SCCN) planned to land a cable in Monterey Bay, California, near the technology center of Silicon Valley. Brett O'Riley, a New Zealander who helped to set up Southern Cross, described to me a trip that he took to the Monterey Bay Aquarium, thinking it would be a good site to meet with cable customers. When he told local museum visitors about Southern Cross's plans to lay an undersea cable, they responded: "You do know that this is one of the United States' most vital marine protected parks? The grand canyon of the ocean?" O'Riley sensed a problem and quickly relocated the Monterey meeting, but the change turned out to be too late. Difficulties in permitting were heightened by the shift to a deregulated telecommunications environment, and despite the fact that Southern Cross gained approval from the Federal Communications Commission and showed that there would be minimal ecological impact, the company responsible for negotiating the landing was never able to secure state permits. As a result, the northern branch of Southern Cross was rerouted through Oregon (one of the project managers notes that this "was dead easy by comparison"), extending the network several hundred miles and increasing its cost by $100 million.[12]

Within the cable industry, California has become well known for its turbulent ecologies. Stories circulate widely about the conservation plans designed for animals along the state's cable routes. When the Japan-U.S. cable intersected the habitat of the endangered Morro shoulderband snail in 2006, the cable company was required to relocate snails along the cable's terrestrial route to twenty meters on either side of it. One manager tells me that his boss was assigned to be a snail watcher, with "Morro Bay Snail Man" written on his hat.[13]

Laughing, he says that after the project was initiated, it turned out that some of the relocated snails were not endangered after all. Another telecommunications expert complains: "Wherever you want to land [in California], there's some rare crab or something that you can't disturb and then you've got to find somewhere else. Then they've also got the ability to try to tax you all the time. Particular counties would you tax you X dollars. Another county would welcome you, but you couldn't disturb the sand dune."[14] One builder gripes that for six months of the year he cannot cross one particular beach in California because it is volleyball season.[15] Some of the engineers suggest that cables' conceptual, and at times physical, proximity to infrastructures such as oil and gas pipelines has contributed to the heightened regulations. Lionel Carter, a marine environmental advisor to the International Cable Protection Committee, suggests that the cable's representation is part of the problem: when marked on a map it appears to be miles wide, whereas in reality it is only 15–20 millimeters thick.[16] It could fit inside these parentheses:

(   ).

Conflicts between telecommunications companies and environmentalists have not been limited to California, however. In Port Alberni, British Columbia, there was a brief clash in 1963 between the Canadian Overseas Telecommunications Corporation and environmentalists when the company cleared forest for cables connecting to the Commonwealth Pacific Cable (COMPAC) system.[17] Cable companies have had to account for frogs in Fiji and move a coral reef in the cable's path in Guam.[18] At some landings, companies hire divers to periodically survey the seafloor to ensure that the cable does not harm the environment (for deeper cables, this is an expensive process that requires remotely operated cameras). In other sites, if a cable company runs its generators during a storm, it may be fined and held accountable for the pollution that it causes. In developing countries or those with limited infrastructure, environmental policies less often constitute such a prominent form of interference. According to one engineer, these governments are afraid to enforce strict regulations: "When it comes to the Pacific islands who are getting their first cable, the governments basically say, 'What do you need?' The last thing they want to do is to scare a cable away."[19] Landing the Honotua cable in Hawai'i took over twenty permits, but landing it in Tahiti required only two.

Although these conflicts with local or state organizations are relatively recent, the cable landing has long been a political pressure point where countries could assert control over international networks that sought to cross their shores. In the United States this was formalized in the 1920s when the Western Union Telegraph Company attempted without government authorization to land a cable at Miami Beach, Florida, to connect with British lines. The cable's superintendent wrote to the U.S. Navy requesting that the company be allowed to proceed since it had already brought together the labor and material at the landing point. Delaying its work, he claimed, would entail "unnecessary expense."[20] However, the Navy blocked the landing until express permission was received. After hearings in the U.S. Senate, a law was passed that firmly gave the president administrative control of cable landings, formalizing what had been established practice prior to that point.[21] Ivan Coggeshall argues that the granting of cable landing licenses "has been the most powerful and most frequently used weapon in the hands of government to control cable companies' rates, taxes, service, [and] national character."[22]

The historical power of the U.S. federal government to defer or delay foreign signals today is today shared with state and local governments in some places. Local residents and organizations are empowered to make demands of the cable companies and benefit from the flow of transoceanic power. In 1990, when installing a cable at San Luis Obispo, AT&T constructed a beach manhole and a parking lot at Sandspit Beach that improved visitors' access to the state park.[23] As these struggles ground flows of power in California, they simultaneously make laying cables to the state from other nations and connecting to the Internet there more difficult. One non-American cable engineer described to me how expensive the permitting process is for his company and the many things that Californian organizations requested in conjunction with its cable landing. He commented: "You can see why California is one of the richest states, demanding all these things being paid for by much poorer countries."[24] Although the engineer's comment simplifies the complex economic relationships between California, the United States, and other countries that seek to interconnect cables there, it illustrates the widespread recognition in the cable industry of the resources it takes to land in California.

The policy governing cable systems, rather than a targeted attack on cable layers, is symptomatic of California's approach to development in general, one in which the historical investment in the state's natural landscape has produced an array of permitting requirements for all kinds of industries. The entanglement of undersea cables in such permitting illustrates how the network's intersection with public and rural environments makes it possible for environmental policy and politics to influence cable development. These conflicts play out in the geography of the network's pressure points, a key site where local and regional actors can generate ripple effects across the system. In is in these locations that cable companies must engage in various strategies of insulation, from disguising warning poles to hiring marine mammal spotters, so they can safely lay cables through turbulent ecologies.

Overall, this power dynamic is in part a product of the lack of diversity in our global networks.[25] If there were more options for connection to the west coast of the Americas, California's landing sites would not be pressure points, and local organizations in the state would have less power to circumscribe the actions of cable companies. At the same time, these strategies of insulation continue to entrench networks in existing local ecologies, and in turn reproduce the lack of network diversity. It remains easier to simply lay cables along the established contours and keep networks in existing landing points, rather than to negotiate unknown geographies and potentially problematic publics. This is especially true in California, where regulatory agencies have been only loosely coordinated and each retains some autonomy. Although there has been recent pressure on governments to better facilitate permitting, it does not seem that restrictions are easing up, and coastal spaces are becoming increasingly regulated.[26] In this environment, the path for action by new companies—or for the development of innovative network geographies—often remains unclear, and the balance of power continues to favor players that have already set up landings and are familiar with local policy.

Do Not Anchor: Cable Reserves in the Coastal Seas

In May 2009 I drive the long, winding coastal road to the cable landing at Morro Bay, California, an important gateway for transpacific communications. Here signals from Asia and Australia are connected to users across North America and transmitted onward to South America and Europe. Along the road, I see a series of posts with faded orange labels that read "Warning: Fiber Optic Cable in This Vicinity" and are constructed to blend in with the natural landscape. When I follow the road down to an oceanside parking lot, I am surprised to see a wooden stand with a nautical chart mounted on it, encased in plastic (figure 4.5). Clearly designated by thick black lines are the routes of undersea cables as they extend over coastal waters and land at the nearby beach. This public display of cable routes is anomalous, given the contemporary concerns about circulating cable information discussed in chapter 3. These markings are present at few landings in the Pacific. Although the prominent visibility of the nautical chart contrasts with the muted visibility of the warning posts, both comprise strategies of insulation devised by telecommunications companies to protect the cable, and each reflects past negotiations within the spatial politics of California.

FIGURE 4.5. Nautical chart posted at Montaña de Oro State Park, Los Osos, California.

Marking cables on nautical charts has long been used as a strategy to keep fishermen and other boaters off the cable route. Historically, the largest threat to cable communication has been from anchors and trawling nets that can sever the cable on the seafloor. The first undersea cable across the English Channel, which took five years to receive approval, was severed shortly after its completion (reports vary from hours to a few days) by a fisherman. A rumor circulated that the fisherman who raised the cable to the surface later exhibited it in Boulogne as a rare seaweed with a gold center.[27] When the Atlantic Cable finally connected Ireland and Newfoundland in 1866, the arrival of telegraph messages in New York was delayed because the last link across the Cabot Strait had been broken by an anchor a few miles offshore.[28] The initial message confirming the working of the Atlantic Cable had to complete its route the old-fashioned way: by boat.

Although an anchor dropped in the wrong place might cut the cable, trawling—a form of fishing that entails dragging nets along the seafloor—is the most disruptive form of interference. Steam trawling, the first major advance in fishing power, was invented in the 1880s, and by the early twentieth century it was regularly disrupting cables in the North Atlantic (figure 4.6).[29] In 1908 the Commercial Cable Company requested that the United States engage in diplomatic correspondence with Great Britain "on account of the

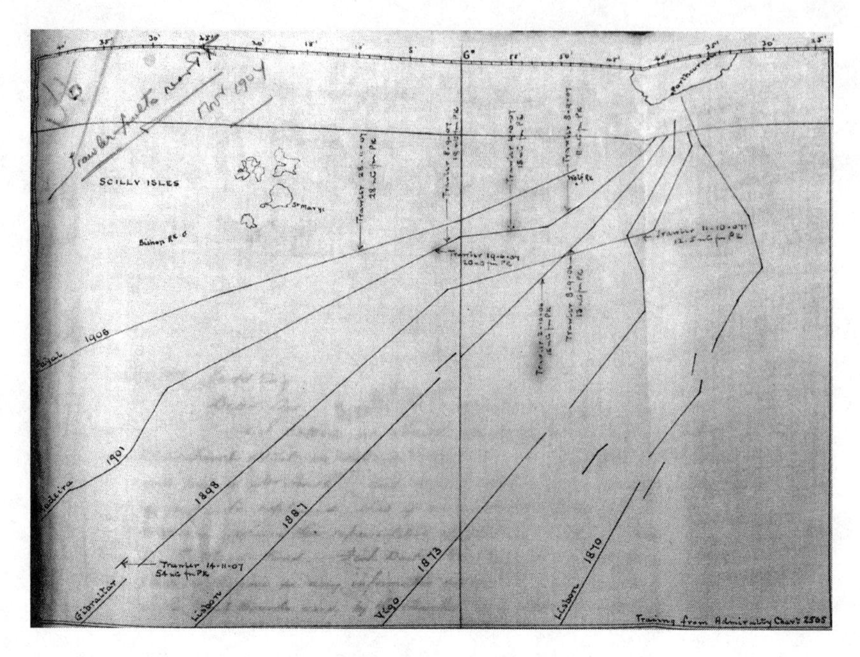

FIGURE 4.6. Chart of trawler damage to Atlantic cables, 1907. © Cable & Wireless Communications 2013, by kind permission of Porthcurno Telegraph Museum.

alleged depredations of English fishing trawlers, who [*sic*] are stated to be constantly destroying the American submarine cables."[30] The company claimed that during the previous three months it had spent over $100,000 in repairing cables off the Irish coast, some of which were broken again only days after being repaired.[31]

Cables had been officially protected by international treaty since 1884, when the Convention for the Protection of Submarine Cables declared it a punishable offense to deliberately damage an undersea cable, but this regulation was difficult to enforce.[32] Cable companies designed a variety of strategies to ensure that their cables were protected. Some cables were buoyed, helping both trawlers and companies locate them. Beacons and lights were installed at bays where prohibited anchorages had been set up. The Eastern Telegraph Company conducted studies and recommended that trawlers' otter boards not be designed with sharp points and protruding parts.[33] The Anglo-American Company suggested that government gunboats be stationed near cables to demonstrate that the cables were protected.[34] The telegraph companies also began to circulate charts to fishermen that indicated where cables had been laid.[35] Some companies from this period pushed for more extensive distribution of charts, but the circulation of this material remained far from widespread. Producing charts was a contested practice, and the decision to release information about a cable's whereabouts was a sensitive one. In an internal letter at the Eastern Telegraph Company about the issue, W. T. Ash wrote: "If we could have a chart printed it may be of some use but on the other hand it may not be good policy to give this information to every person, whether English or Foreign."[36] Other correspondence cited "the undesirability of publishing these charts broadcast."[37] The 1884 Convention also stated that if fishing gear or an anchor was sacrificed to protect the cable, the cable companies should compensate the owner, yet this too was difficult to enforce. The German-Atlantic Telegraph Company, facing many "frivolous claims" from fishermen for lost gear on cables, requested that the fishermen buoy their gear when they dropped it.[38]

In response to the increasing cable disruptions, in 1908 a UK Interdepartmental Committee on Cable Protection met with representatives from eleven different cable companies and presented its findings to Parliament.[39] The companies requested that trawling be prohibited in the cable area, a strategy of insulation that would block off an expanse of 3,400 nautical miles. However, the committee refused to recommend the establishment of such a reserve, commenting that "it would be an unjustifiable interference with a business which represents a capital of several millions, gives employment to a body of hardy and industrious men, and supplies a substantial part of the food of the people."[40] The committee claimed that shutting off this area would not only limit fishing capacity but would also provide inadequate protection for cables, as "enforcement would involve police measures of a difficult and costly nature."[41] Moreover, there was the problem of nationality: setting up laws to regulate British fishermen would not be enough, since fishermen from other countries could continue to trawl the area. If there were such a proscription, it would have to apply to all vessels from all countries. The committee also determined that charts might not be useful in practice since navigational equipment in this period could not precisely determine a boat's position when out of sight of land. The committee decided that the best thing to do was to help the fishermen make their gear safe for cables. It suggested modifications—heavier cables for the company and better gear for trawlers—along with a system of government inspection. In 1908 policy was on the side of fishing, which was seen as a necessary industry with a prior right to the seafloor; the British government refused to insulate cable traffic by limiting existing circulations.

During the next fifty years, cable companies continued to develop strategies of insulation: publishing the locations of "no anchorage" zones, notices to mariners, and cable charts. These were not coordinated internationally, and government decisions continued to favor fishing interests over those of the cable companies. For example, when cable companies in New Zealand requested to have specified conditions about telegraph cable protection printed on the back of fishing licenses, their request was denied. They were instead advised by the mercantile marine to put out a circular to trawlers letting them know that if they wanted to avoid cables, they could ask the Pacific Cable Board about the cables' locations. Only if this did not work would the government consider establishing prohibited areas, but this option was "inadvisable," as it would greatly inconvenience fishermen.[42] After an Auckland-Suva cable was damaged in 1949, the New Zealand Marine Department determined that there was "insufficient evidence" to take any action against the trawler.[43] In a letter discussing the damage, the department suggested that the cable companies were partly at fault: if cables were laid with enough slack, they would fall "flat on the sea bed and so present little hazard to trawls."[44]

Cable companies appealed to fishermen directly, sometimes meeting with associations of fishing vessel owners and attending their exhibitions. For example, in 1922 the Eastern Telegraph Company set up a large exhibit showcasing cable history and technology at the Deep Sea Fishing Exhibition in London, attempting to forge connections between the industries. The company magazine noted that the "Fishing Industry is closely allied to our own and prior to the Associated Companies coming into the open, the fisher folk regarded us as their enemies."[45] The company attempted to articulate the importance of cables to the fishing industry: "the world must have cables, for without cables you cannot sell your fish."[46] At other times, companies tried to emphasize how dangerous cables were. One notice read: "HIGH voltages are, or may be, fed into certain submarine cables; SERIOUS risk exists of loss of life due to electric shock, or at least severe burns, if any attempt to cut the cable is made."[47] This problem remained far from resolved, as cables continued to be snagged by trawlers' nets throughout the first half of the twentieth century.

The advent of higher capacity coaxial systems and the increasing emphasis on security in the 1950s amplified the need for strategies of insulation. Cable & Wireless estimated that by this point hundreds of thousands of pounds were being spent each year on repairs of trawler damage; the company reported a fishing interruption almost every other week.[48] To coordinate their protection efforts, cable companies formed the Cable Damage Committee in 1958. One of its early activities was to coordinate the production and distribution of cable warning charts in the problem area of the North Atlantic. This two-year process required gaining agreement from the governments of a dozen countries. Arthur Harris, a Cable & Wireless employee, comments that cable companies struggled with "limitations imposed by Government authorities" on the publication of cable routes.[49] He reflects that the cable routes "were in fact a carefully guarded secret" up to this point, and that the only charts produced showed cable paths "in a general way which if the fishermen were to take them seriously would have stopped fishing altogether."[50] At this historical turning point, the Cable Damage Committee was able to facilitate the sharing of material in a wide range of discursive contexts.[51] It made tape recordings in nine languages that asked fishermen not to trawl the cable routes (these were then broadcast over the ocean) and published advertisements (including cartoons) in fishing news media. It sent out cable ships to patrol the northwest Atlantic. Some companies even distributed calendars with girlie pictures, which urged fishermen not to pull up cables if they were snagged.[52] The committee and the cable companies also produced films for fishermen, such as *The Catch That Nobody Wants* (Great Britain, Rene Basilico, 1973).[53]

During this period, the committee's efforts led to the passage of a number of laws that enabled the declaration of protected areas within territorial waters for cables.[54] In addition, cable companies also developed a new strategy to protect cables from fishermen: burying the cables under the seabed. Bell Telephone Laboratories designed a fourteen-ton sea plow, first used in 1967, that could bury a cable twenty-eight inches deep.[55] Later, techniques for directional drilling (which extended a conduit for the cable horizontally under the seabed) and for rock dumping (in which rocks were placed over the cable) insulated the network by moving it even deeper beneath the seafloor, though cables still risked being unearthed by seabed movements. These processes also served to further remove the cable from public view. There was no longer a visible cable landing, as there was nothing to see when the cable reached shore. One engineer told me: "It took away the spectacle, but it also takes away some of the security risk, because people don't even know it's there. It's buried and concrete encased."[56]

Even in their reorganization under postwar security measures, companies could not fully insulate the cable landing point. The new protection acts remained difficult to enforce, as the burden of evidence still fell on the cable company. Although there were successful court cases, these were "few in comparison with the total number of cables that [were] broken."[57] Philip Kelly's discussion of security of the TAT-1 telephone cable testifies to this: "The reliability of everything used in the submarine sections . . . was paramount. Manufacture under totally dust free and hygienic facilities added a measure of 'star wars' type clothing and security. All was highlighted to good effect by the fact that we had no failures of the repeaters, but there were of course a few cable failures due to fishing activities off the Newfoundland fishing banks. Thanks to the decision to locate a cable ship already loaded with spare cable and repeaters off Newfoundland, any failure was of relatively short duration."[58] Every step of the process was brought under control, including using only proven technology and "star wars" security, but Kelly still writes that "of course" there were still failures due to fishing, and the only real way to insulate one's cables was to locate repair ships nearby.

Although the coastal ocean, such as California's landing point, had long been the most turbulent environment for cable companies to cross, conflicts—and the development of modes of insulation—escalated as marine spaces became more impacted in the fiber-optic era. Companies and fishermen faced new of forms of interference in their attempts to use the landing point, from offshore installations of oil rigs and wind turbines to marine sanctuaries, fish farming, and heightened shipping traffic. Breaks of cables due to fishing remain frequent in Southeast Asia, where the water is shallower, and at cable hubs that have historically been port cities, such as Hong Kong and Singapore. In China fishermen have nets that are held down by large weights, including materials such as train wheels, that can slice through a cable. At these sites, one cable engineer tells me they have to bury the cable up to ten meters deep, so deep that it would be difficult to retrieve if anything happened to it.[59] Other breaks remain beyond the control of ships, which by no fault of their own might be dragged by a storm across a cable route.

Much more now relies on a single cable—in some places, an entire country's connectivity. Because the Internet has been reframed globally as a critical resource to the information economy, the balance of power has shifted to the cable companies. With cable repair costing millions of dollars, it is in a country's economic interest to protect these systems. In the last decade, the International Cable Protection Committee (formerly the Cable Damage Committee) and cable companies successfully pushed for the establishment of new cable protection zones. New Zealand's government concluded that its law was not achieving its purpose: underwater surveys of existing cables reported over a hundred anchors, nets, and other objects lying near or in contact with cables.[60] The updated Submarine Cables and Pipelines Protection Act of 1996 greatly increased penalties (from $1,000 to $250,000 for breaking or damaging a cable and from $5,000 to $100,000 for commercial operators fishing in a restricted area) and allowed a broader range of evidence to be taken into account, shifting the burden of proof away from the cable companies to the boaters. The act also granted new powers to enforcement agencies, permitting them to obtain identification and documents from a ship. People who fished in protection zones now could be prosecuted even if they did not damage a cable. Automatic Identification Systems, technologies that record the location of ships and enable remote surveillance, also allow some companies to determine what ships are responsible for faults.[61] Together this made it easier to prosecute fishermen and other offenders.[62] To establish this process, dreamed up by cable companies a century earlier and facilitated by cable protection acts, took the better part of a decade and much lobbying by the International Cable Protection Committee with fishermen and governments.[63]

Well-enforced cable protection zones afford insulation to new cables: they are sites where signals are routed more smoothly. In Australia protection areas were set up around existing cables, and much of the country's international traffic now exits through two zones in Sydney. Some people believe this is sufficient, but others believe that more diversity is needed: if both of these zones were to be disrupted, the country's network traffic would have few other exits. Brisbane is one option for a cable landing, but there is no protection zone there. Of course, companies do not need a protection zone to land a cable, but without one they would have to apply for a non–protection zone landing license with a fee of $8,176 and an expert consultancy charge of $25,000—which is significantly more expensive than a protection zone permit costing $2,215 (all figures in this and the next sentence are in Australian dollars). To apply to establish a new zone, companies would have to pay a fee of $162,000.[64] As one prospective cable builder observes, although the government would prefer that a cable be laid before it establishes a protection zone, it is economically advantageous for the company if the zone comes first.[65] Any company that went through the process of establishing a new zone would have to pay for it, and the company's competitors would benefit from its investment. The costs are not "prohibitive," but the cable builder tells me that there is no commercial advantage in going into Brisbane, especially since cables there will not receive the instant federal protection that Sydney's cables do. As a result, cable companies remain at a standstill. The existing strategies of insulation give Sydney traction, discourage diversification, and ultimately leave Australia's information traffic vulnerable.

Like the negotiations with environmentalists, such imbrications in the coastal ocean are both historically and culturally specific.[66] Australia and New Zealand have pioneered this form of cable protection; in contrast, most countries have not devised a formal approach to insulating the landing point. In California, and especially around the Morro Bay cable landing, fishermen still have significant power and continue to shape the contours of the cablescape.[67] One station manager told me, "If it were up to those fishermen, we'd never lay another cable to the California coast," and some in the industry suggest that cables may continue to be routed through Oregon because the fishermen there are "a bit more accommodating."[68] The approach to insulating one's signals in the United States has not entailed a rigid enforcement of protection zones. Instead, it has led to the creation of independent committees, such as the Central California Joint Cable/Fisheries Liaison Committee, to act as liaisons between fishermen and cable industry. The committee not only distributes charts and other important information to fishermen, but it also offers them grants and other financial resources funded by the cable companies.

Rather than attempting to keep the fishermen out, such measures channel resources to fishermen to keep them invested in the cable's sustained operation, linking them to the network that helps support their own mobility. One California Coastal Commission report lists some of the mitigation measures requested of applicant cable companies:

- Fund a Committee/Liaison office to the amount of $50,000 annually per cable company with funds in excess of $150,000 being transferred to the Commercial Fishing Industry Improvement Fund. . . .
- Annually deposit $100,000 per project in a special fund for the enhancement of commercial fisheries and the commercial fishing industry and support facilities. . . .
- Pay $500 to each licensed fisherman who signs the Independent Agreement for use in upgrading communication and navigation equipment.[69]

In Japan, Korea, and the United States, money is regularly transferred to the fishing industry so that companies can lay cables, but this practice also goes on informally. One cable engineer told me that although his company initially had "fishermen problems" around Southeast Asia–Middle East–Western Europe (SEA-ME-WE) 3, after talking to the fishermen and buying a few licenses from them, the problem was solved: "So they got money, so we got rid of a point of aggravation. At the end of the day, we became sensitive to their concerns."[70] As in the early days of the telegraph era, rumors circulate in the cable industry about the "opportunistic goals" of the fishermen. One cable engineer recounts a story about boaters who would deliberately fish around a cable: "If they made enough of a nuisance of themselves," he tells me, "they would be paid off."[71] Such negotiations, involving either grants or informal payoffs, represent strategies of interconnection that tie together the companies' vision of the ocean as a place for fixed technological development and the fishermen's needs for unrestricted mobility.

The production and circulation of cable maps—a strategy of insulation that took more than a century to formalize—tell fishermen where the cables lie and how to avoid them, but the maps also inadvertently create public documents of all of the cable landing points, even if they do not show the precise shore ends.

In California, as cable routes are negotiated through the immense documentation required in the EIR and made visible in cable charts, telecommunications companies end up disseminating to the public material on the cable's location. Due to the public interest in preserving these lands as natural spaces and fishing grounds, many reports have even been published online. This has resulted in the creation of an archive of visible evidence of cable infrastructure—about both the network and its physical sites, such as the map posted at Morro Bay (figure 4.5)—that otherwise would have been deliberately hidden. For the cable industry, this is a price to pay for the ability to insulate their traffic, but for military cables in the United States, the opposite approach has been taken: these cables are kept hidden and unmarked, disruptions in them are simply allowed to occur, and repairs are made in them after they break. A representative from the Naval Seafloor Cable Protection Office tells me that although "for industry the logical answer has always been, 'we need to get the information out because that is our greatest source of faults,'" the U.S. Navy will "rarely go to the fisherman and say please don't go here. Because fishermen go where fishermen go."[72] In some cases it is simply better to deal with the consequences of a break than to give up sensitive information about its location. Strategies of insulation must balance the advantage of publication for protection and the need to keep infrastructures hidden for protection.

Throughout history, as they have extended cables through the coastal seas, companies, governments, and militaries have come into contact with fishermen who generate friction for cable traffic. In the early telegraph era, cable companies struggled to protect their systems in a geopolitical context that favored fishing interests, but through the coaxial and fiber-optic eras the balance of power slowly migrated toward the cable companies. This has made it easier for them to set up strategies of insulation, from the widespread establishment of cable charts in the 1950s to the recent enactment of cable protection zones. These strategies, like those meant to stabilize signals as they cross the California coast, are always designed in relation to local cultural ecologies at the system's pressure points and, in turn, help solidify cable routing along existing contours. These are sites where the local movements of boaters and fishermen, like the California environmentalists, can produce large-scale effects on network operations.

Do Not Cross: Farrington Highway, O'ahu

On O'ahu's west shore, I stand among a group of beachgoers in a parking lot, snapping pictures of a manhole. Tourists are less common here than in downtown Honolulu, and the fact that my camera is pointed at the ground rather than the horizon further sets me apart from the crowd. A bearded man in his thirties named John approaches me and asks what I am looking at.[73] I tell him that this is a cable landing point, a place where undersea cables come in from across the Pacific. "You're kidding," John says, pointing to the manhole. "You mean I could crawl in there and get to Guam?" I hold my hands up, making a circle of my thumb and index finger to approximate the size of a nickel, the width of an unarmored cable. "If you were this small," I tell him. I take out a set of maps and show him the locations of cable landings that I have gleaned from environmental permits and nautical charts. "I've never heard of this. You should write up an article in the *Honolulu Advertiser*," he tells me. "The people here would want to know about it." We walk up the beach toward the cable station, but I am reluctant to go closer. "I don't want to get in trouble," I tell him, but he assures me that no one will mess with him because he is a local. He does not share my hesitance at approaching secured sites.

John offers to take me to the places I want to visit. Since he has lived on the west shore for most of his life, he knows the area well. Over the next few days, we travel up and down Farrington Highway, exchanging our knowledge of local geography. He points out the tent cities where houseless people live off of public infrastructure constructed for beachgoers, the hills where his grandmother resided before the military took over the land and moved her down to the coast, and the large valleys where one can see weapons testing. I point out the manholes where the cables are pulled up, the barely noticeable difference in the concrete road that indicates buried infrastructure, and the cable station embedded in military lands. As we drive up the coast, I tell him that his neighborhood is a critical hub in transpacific networking. All of the island's coaxial and fiber-optic cables since the 1960s have terminated at three cable stations along this thirty-kilometer stretch. He knows little about this: aside from the manholes and the signs warning residents not to dig in the area, there is little to signify the concentration of cable systems.

It turns out that these two spatial formations, the technical geography of cables and the cultural geography of militarization and economic deprivation, are closely intertwined. The conflict between them plays out on the area's one main road, Farrington Highway. Highways are often thought of as nonplaces, spaces we simply pass through and do not reside in.[74] However, Farrington Highway is a central route for many on the west shore: the military, industries, and local communities. The highway was established next to the original railway mainline that linked military establishments on the south and west shores of O'ahu prior to and during World War II. In relation to the island's

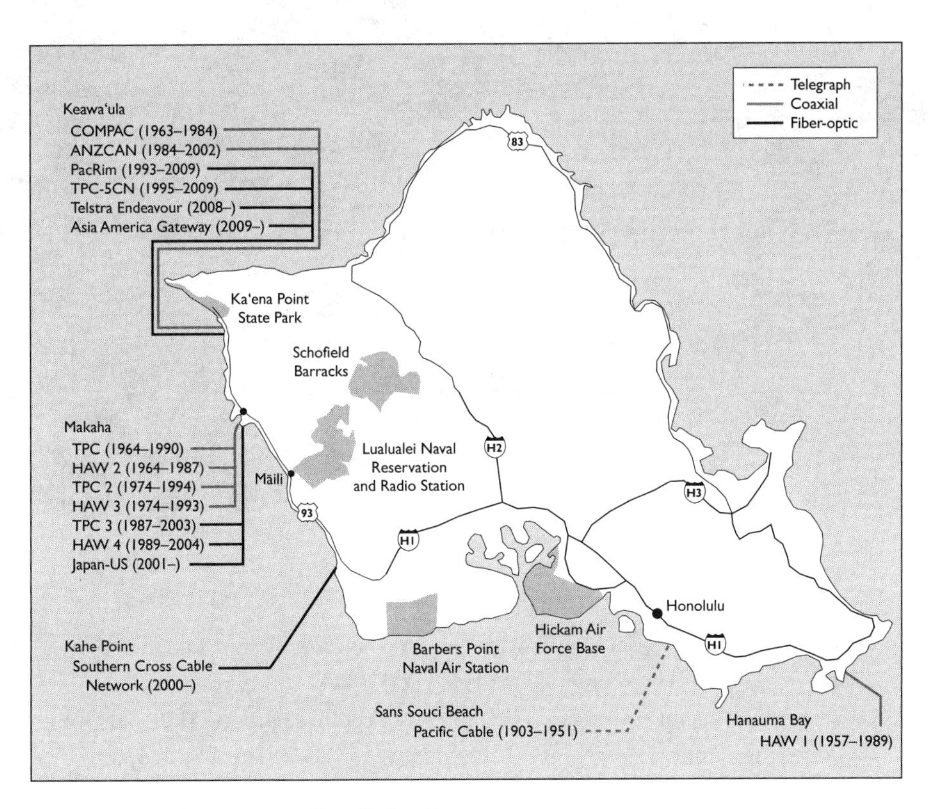

MAP 4.2. O'ahu's transpacific cable landings.

broader transportation geography, Farrington Highway is marginal. A much larger freeway, the H1, runs along the southern shore of the island, connecting its most densely populated areas. The road then continues along O'ahu's east coast, around the island to the North Shore and back down through the center of O'ahu, running directly into Honolulu. It forms a complete loop, which—like the cable circuits that land here—is a robust infrastructure as it allows at least two possible routes to any given location (map 4.2). In contrast, the smaller Farrington Highway does not connect to the North Shore, ending instead at Ka'ena Point State Park. Hugging the hills, it has no parallel paths; there is no way to cut through to the center of the island. John tells me that on the west shore, there is "only one way in and one way out." Farrington Highway, in other words, is a dead end.

Despite this, or perhaps because of it, Farrington Highway is also one of the primary places where social life is constituted on the west shore. Cars slowly cruise up and down, and when people spot each other on the side of the road, they honk or wave. When I ask John how long it takes to get around O'ahu, he

FIGURE 4.7. Farrington Highway, O'ahu.

answers that it takes ten hours, even though when I drive it, it take less than half of that time. His reply tells me that on the west shore, roads are places where people spend time. Most commerce is structured around the road. We visit beaches immediately from the roadside. When the road is blocked, traffic becomes unmanageable, and everyone gets upset. The spatial organization of Farrington Highway has developed in this way not only because of the island's geological formation (steep hills rise beside it) but because the military occupies and prevents access to many inland areas. The Air Force Ka'ena Point Satellite Tracking Station, the Makua Military Reservation, the Lualualei Naval Reservation, and the Schofield Barracks separate Farrington Highway from the rest of the island.

Although Farrington Highway is the only route connecting points along the west shore, it also acts as a dividing line. Inland, on the *mauka* side of the road, are weapons testing sites and military installations.[75] Native Hawaiians, the Kanaka Maoli, have been forcibly displaced by the military from many of these locations, including Makua Valley, a sacred site that is linked to the creation of the earth. John tells me that his grandmother has been removed from this valley, and that he has not been able to visit the area. There have been numerous protests at these sites calling for the return of this land. On the *makai* side of Farrington Highway, all along the beach, are the tent cities where houseless people live.[76] The trash from these dwellings is thrown alongside the road, often on the *mauka* side, next to the military installations (figure 4.7). I am repeatedly told that this is not the "safest" part of O'ahu, and I hear stories about surfers from outside the area being harassed here. Although the military has occupied these mountains, Hawaiians have occupied the shoreline, making it difficult for outsiders to approach the ocean as they do in the rest of Oʻahu.

Just as at the landings in California, Australia, and New Zealand, cables here are implicated in local spatial politics. They have been laid to the west shore in part because of the military occupation, which not only provides security but also has historically been one of the largest customers of cable capacity. Beyond this, the friction-free operation of the cable station directly depends on its connection with other friction-free local infrastructures, especially the road. The durability of these infrastructures, as well as the local politics with which they are intertwined, can easily affect the operation of the cable station. Farrington Highway is critical for the workers' daily commute to the stations and for the transportation of equipment and other goods; it facilitates the flows of the human and physical resources necessary to keep the station running. As David Morley has observed, communication in many ways still depends on physical transportation.[77] At one point, a huge storm washed out part of the road. With only one way in and one way out, employees could not reach the cable station. The two men who were there at the time had to remain until transportation was reestablished. The only other option would have been flying their replacements in by helicopter, a process the cable station manager had to consider. Other natural disasters that occur in the area include brush fires and landslides. At one point, firefighters literally stood on the cable station at Keawaʻula and fought back the flames. Support to cope with these disasters likewise depends on the friction-free operation of the road.

The turbulence of Farrington Highway originates both in the natural disasters that could easily disrupt it and in the communities whose members also depend on it and have a stake in its operation. In 1999 AT&T undertook to reconnect its two stations along Farrington Highway. Given the narrow space beside the road (there is often little room between it and the encroachments on either side), this included digging up much of the land and delaying traffic. This disruption in the daily lives of local people caused an outcry, and the cable station manager tells me that some of the residents came to think of AT&T as the "enemy."[78] Just as the memory of the incident has lingered in the minds of some residents of the west shore, the cable route remains visually signified by lines in the concrete and faded orange "Warning: Buried Fiber Optic Cable" posts planted alongside the road. On Oʻahu, residents have reacted every time a new cable is proposed, and the beach and road—spaces that are integral to local cultural practices—have to be trenched in order to extend the cable to the ocean. Since AT&T cut its community relations position in O'ahu, the cable station manager spends much of his time addressing the claims of residents in the towns along Farrington Highway, a site that he also describes as "one way in and one way out." People here don't like anything that disturbs the status quo, he explains. As the company does for the California fishermen, AT&T funnels resources to the community in return for the ability to land its cable on the west shore, a process negotiated via neighborhood boards instead of liaison committees. For example, at one meeting, "Saylors moved . . . that the [Waianae Neighborhood Board] support the project of the submarine fiber optic cables proposed by AT&T and Telstra with the benefits package from AT&T that includes at least two $5,000 scholarships for our two local high schools, Nanakuli and Wai'anae, and the possibility of land benefits and that Telstra also is looking to see if they can make some kind of donation to the land in the area of Keawa'ula/Yokohama. These scholarships will be awarded annually and will be made available as long as AT&T is using the cables at Keawa'ula/Yokohama."[79] Farrington Highway is a site of struggle between these two modes of spatial organization, and, to a large degree, the struggle between them has been aggravated by the visibility of cable laying. Future cable projects, if any are permitted, will likely use directional drilling, which will keep the cable out of the way of traffic. As it ensures cable work will not disrupt the road, this strategy of insulation preserves existing social ecologies.

Decisions about trenching through or boring under the road might seem to mean little in comparison to the establishment of multimillion-dollar cable networks, but creating strategies of insulation such as routing cables in nonintrusive ways are an integral part of the industry. Companies cannot afford to assume that the environments through which they lay cables are friction-free. As recounted in chapter 1, during the dot-com bubble, the speed of cable development meant that less time was spent on these sorts of strategies during infrastructure planning. In 2002 Tyco Telecommunications built a huge $75-million cable landing facility in the middle of the town of Ma'ili, Hawai'i, with a landing point on the town's beach, planning to integrate the location into its global network. The resistance of this community—which, as in California, was enacted through the permitting process—reportedly hastened the network's failure. The cable was never landed, and the building was put up for sale. A news article written two years afterward notes that the price had dropped to $5 million.[80] The community has since considered purchasing it for a homeless services center or transforming it into a community college, but the building is still too expensive.[81] Whether in Hawai'i or California, or at any other landing point, failing to adequately address such turbulent ecologies can delay network development.

Although the Maʻili cable station was located in the center of town, the landing facility constructed for the Southern Cross network in 2000 took an opposite tack in negotiating visibility and used marginal spaces to keep itself deliberately out of view, especially from people in the local community. The Kahe Point station is outside of town, and the building is tucked under a cliff and hidden behind trees. It was built next to the shore, a location that meant the cable would not have to be routed a long distance or through inhabited spaces. The building did not have an address until the absence of one became a problem for visitors and deliveries. Indeed, the station and the cable route is so invisible to the community that at one point the local police came to the building to inquire about its function. Unlike the routes maintained by AT&T, the cable here is not marked by "Do Not Dig" signs. Instead, the manager routinely goes out to check the cable route and the perimeter of the station. The company's strategy, delineated in the previous chapter, is to remain as low-key as possible, minimizing infrastructural visibility and contact with the community.

On Oʻahu's west shore, cable companies' strategies of interconnection and insulation—from funding local schools to maintaining only a marginal presence—reorganize social space to maintain signal traffic. As on the California coast, insulating the cable entails negotiating its visibility to local publics; in the case of Oʻahu, the community is attuned to the struggles for land rights and culturally connected via the road that the companies need to cross. As a result, the development of transpacific cables on Oʻahu's west shore, a pressure point in transpacific networks, has taken place in relation to the existing political contours of Farrington Highway. The territorial occupations of the military facilitated the establishment of the cable stations along this narrow stretch of road; the territorial occupations of the community and its members' stake in the local transportation systems have served to check the expansion of global Internet infrastructure. Like the development of environmental permits and nautical charts, this is not an insignificant expense in the continuing operation of cable infrastructure.

Diverse spatial tensions have emerged in different parts of the world—from potential conflicts with Maori people in New Zealand to contested sites in Bamfield, Canada—but these tensions are able to escalate and affect cable development in Hawaiʻi and California in part because of American bureaucratic processes that empower local and state organizations.[82] In the United States, the president's National Security Telecommunications Advisory Committee has recommended that the government reduce such forms of interference to speed up cable development. The committee recommends improving this regulatory environment, "currently subject to a myriad of regulations and rules regarding placement and service," to decrease the vulnerability of the network's pressure points.[83] The cable industry has also pushed for more cohesive national regulation of cable landings and more informed governmental involvement, which in turn would facilitate the establishment of intercontinental networks and ensure that the existing infrastructure will be less precarious.

In some countries, the state and national policies that regulate network development already give telecommunications companies these benefits. In Australia, carriers laying underground cables are given immunity to state and territory laws, and as a result cable companies are able to bypass long, expensive, and heterogeneous engagements with local governments.[84] This makes it far easier for other countries to connect with Australia. One cable engineer describes his company's building process in the country: "We didn't divert around obstacles, like people's driveways or water pipes . . . just went straight through them. And then a team of other people followed to replace water pipes, reinstate driveways."[85] In this model, power shifts from the hands of communities—which in the United States are able to affect global infrastructure via local policies—into the hands of telecommunications companies. One cable operations manager tells me that, in turn, his company tries to do community outreach and make what they are doing visible to reduce any potential conflict, since "the biggest problems come when people don't understand what those powers are and think we're being cowboys."[86] At one point, his company hosted a barbeque and asked residents to come see why it had occupied their park for several months. Another cable engineer recounts that "a lot of [public relations] went into it to make sure the objections and problems were minimized . . . you make it into a social event, you get people involved, the television, the radio stations, the community and the local government."[87] By getting local communities to be invested in the cable landing, the company engaged in an attempt at interconnection, despite the fact that it already had the right to develop in the area.

Telecommunications workers in the Pacific often attribute friction at the cable landing point to the self-interested motivations of opportunistic actors (one cable builder told me that "it's a bit of a litigious environment and a good opportunity for lawyers to get into a bit of a spat").[88] Another way to understand these coastal conflicts—which have accelerated in the fiber-optic period—is to see them as assertions of territorial agency in the wake of privatization and deregulation, processes that stripped control from nationally backed companies. The domestic companies that had arranged cable landings since the 1950s had typically been nationally affiliated monopolies with a history of working with the state bureaucracy. As this system was broken up, there was a barrage of uncoordinated attempts to penetrate national borders, especially by less experienced companies attempting to lay cables in new zones. Telecommunications companies thus came into contact with these environments at unprecedented speed, without having negotiated strategies of insulation or interconnection. The heightened opposition of fishermen, the protests of environmentalists, and the contestations of communities on the west shore of Hawai'i occurred in response to this lack of order, the breakdown of state control, and the encounter with new forms of transnational development, rather than as a direct and specific opposition to cable systems. In seeking to preserve existing modes of spatial organization, these actors generated turbulent ecologies for the cable layers.

Limits to Visibility

A few days after a typhoon swept through the area, I take a bumpy cab ride three hours outside Manila, the last node on the original transpacific cable route to Nasugbu on the west coast of the Philippines. For the United States I have charts and environmental documents made public in part due to local actors' entanglement with state agencies, but documentation of the landing sites in the Philippines is not available. I cannot generate close-up satellite views in Google Earth. After several failed attempts (the language barrier means that I keep getting directed toward cell phone towers when I ask about Globe Telecom), I finally show the driver a picture of the station from the publicity website, which he then shows to local residents. It is only through this visual communication that I end up at the cable station. The Philippine Long Distance Telephone Company and Globe Telecom stations are located next to each other, on the same street in the center of the same small town. The publicity photos I have seen do not match the view from the street, where large walls block my view. Neither structure has been disrupted by the typhoon. All I can glimpse beyond the walls is a basketball hoop at the top of one station and a satellite dish when the gate opens to the other. A different sort of infrastructural visibility is in operation here: although many stations blend into their surrounding environments, whether an industrial landscape or a nature preserve, in the Philippines the station stands out, but my vision is directly impeded.

As a U.S. citizen, I have been cleared to enter cable stations on American soil, and as a researcher, I have the social capital required to justify such visits. However, here I am unable to gain entry. Instead, I walk toward the ocean along the path where I assume the cable is laid, though if there are markers that make its route visible, I cannot see or recognize them. Instead, I am the

FIGURE 4.8. Nasugbu cable landing, the Philippines.

visible one: one of the few women on the street, young, white, American, holding a camera, and looking for something to photograph. The ocean surges up an alley that might otherwise provide access to the beach (figure 4.8). A young man waves for me to come through the huts that border the alley. He moves a fence aside, and I duck in. Clothes dangle from the ceiling, a chicken runs by, and trash is swept up through the house by the ocean. This last stop, near the original Commercial Pacific Cable route, is a reminder that my ability to navigate and even perceive cable infrastructure is contingent on my nationality, racial and gendered assimilations, and cultural familiarity with how local geographies have been organized.

Although the cable landing point is the place where network infrastructure surfaces and becomes intelligible in social space, not everyone has the ability to affect the composition of these spaces. In the United States, state and local governmental regulations give communities along the cable route the capacity to intervene in the development of cable networks. This does not hold true in places like Australia, where telecommunications companies are able to bypass local regulations. In some places, one can detect the cable system through a range of visual markers—such as manholes, conduits, flags, posts, and spray paint on the ground—but not everyone has the ability or interest to do so. Cable companies have long known this, and the cases in this chapter illustrate how they have consciously managed the visibility of cables since the nineteenth century. In the state parks of California, cable companies' strategies of insulation entail hiding the cable. In California's coastal waters, companies have long fought to make the cable visible and intelligible to local fishermen. Because they intersect public spaces, the development of strategies of insulation and interconnection at the network's landing points always involve the manipulation of such visibilities and publics.

In telling a set of nodal narratives about cable landing points, this chapter has offered a new way to look at our networks. We might think of landing points not as environments that are made irrelevant by signal traffic, but as places where global cable systems encounter friction and experience tension with human and nonhuman inhabitants of aquatic and coastal space. Landing points are an interface between technological networks and the local cultural practices they seem to bypass, a bridge linking previous and future circulations. They are also the pressure points of our global nervous system, where only a small amount of resistance can have significant impacts. As a result, relatively small-scale conflicts between commercial buildup and the preservation of wildlife, coastal property values, and local economies in California have shaped transpacific cable development. In Hawai'i the politics of one road have made it difficult to land international systems along the coast. The cable landing point is a space where local actors have global agency, where the political power of the telecommunications companies is contested and diffused, and where network environments are made turbulent.

Paying attention to these turbulent ecologies lends a critical layer to our understanding of global networks, one that can help account for the contemporary struggles of the cable industry. On one hand, these ecologies illustrate the continued significance of humans and nonhumans in aquatic and coastal environments as well as the failure of cable companies to fully transform such environments into friction-free surfaces. There are still hundreds of cable faults around the world that result from fishing and anchors every year, and there are long lag times for permitting and development. The strategies of insulation created to stabilize the movement of signal traffic through these environments continue to cost cable companies money, time, and labor. Indeed, one of the reasons that cable maintenance is so expensive is the cost of keeping repair ships on standby around the world in case of a break. Together, these investments at the undersea network's landing points mean that cable routes remain resistant to change, and information flow, funneled through such pressure points, continues to be affected by the materiality of the world it traverses.

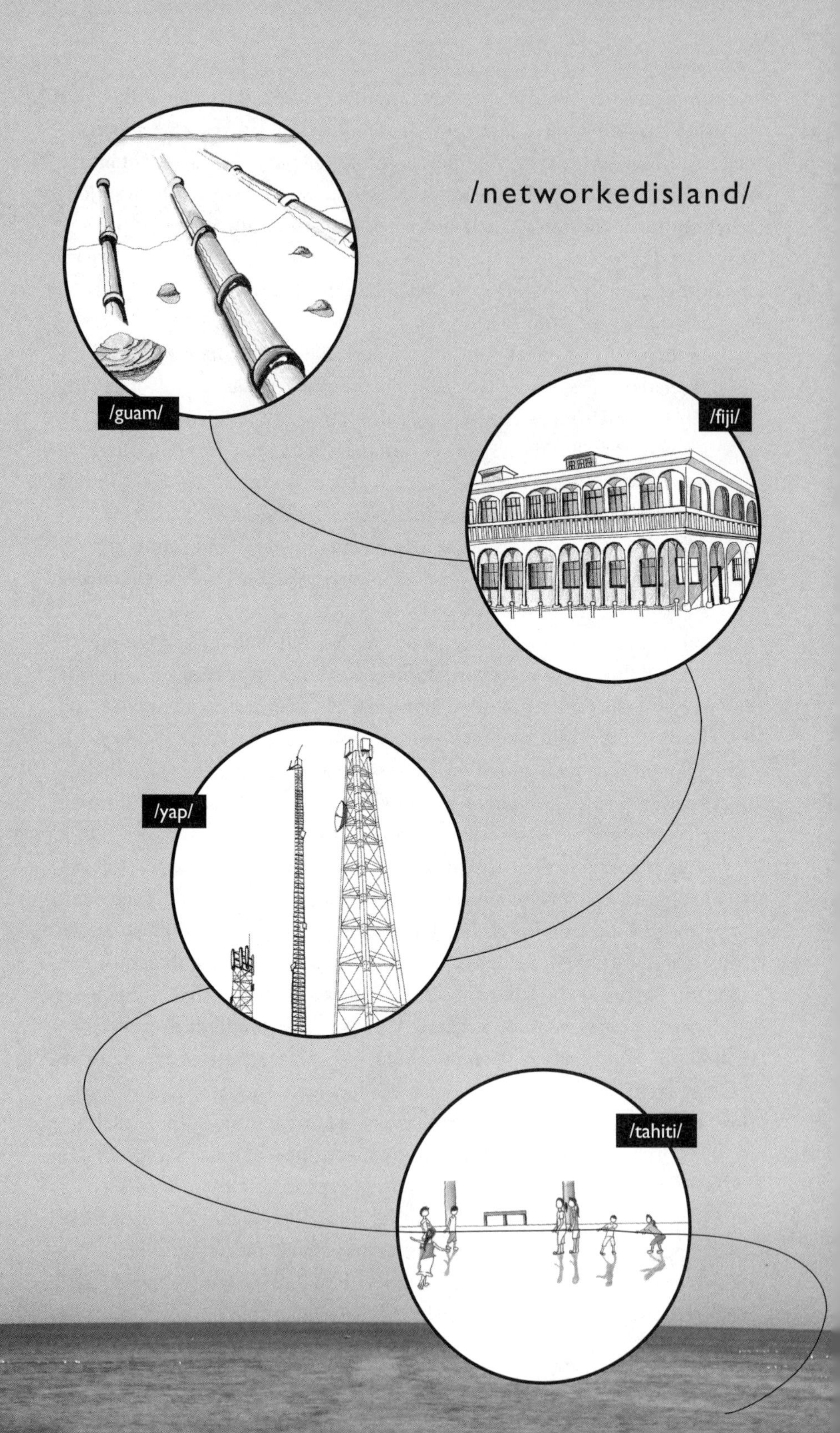
/networkedisland/
/guam/
/fiji/
/yap/
/tahiti/

FIVE

"The Cable Service is rich in islands . . . a complete estate within a very definite boundary, which from any point of vantage comes entirely into view, they give to the beholder a pleasant feeling of undisputed sovereignty."—"Small Islands"

As the telegraph system was extended around the world, networked islands from Bermuda to Vancouver Island and Hong Kong became critical gateways to empire.[1] The island was an apt geography for the colonial cable network as it provided insularity: it was physically separated from the mainland and its potentially threatening populations, and it could be more easily secured by military forces (which also needed communication services during times of war). Cable stations afforded a kind of conceptual sovereignty and perceived control over these oceanic "stepping-stones."[2] Today, islands remain important to Internet communication even though fiber-optic cables can span the Pacific Ocean in a single hop. More than half of the nodes in our undersea cable network (366 out of 685) are still located on islands. There are more landing points on the islands of Hong Kong and Taiwan (eight) than on the entire mainland of China. Fifteen cable systems connect Singapore to the network; only thirteen are linked to the rest of mainland Southeast Asia. Mainlands often depend on islands for their communication: all of Australia's cable traffic passes through islands, which must remain above water and politically stable, before it reaches any other continent. The previous chapters delineated the geography of the cable station (the gateway to undersea networks) and the landing point (where cables extend through coastal territories); this chapter examines how islands have been critical sites for cable networks. Recasting our global network as a network of islands shifts our imagination of media infrastructure to take into account its coastal and aquatic encounters and the significance of locations such as Hawai'i and Guam in mediating intercontinental traffic.

In contemporary American culture, islands and networks appear to be mutually exclusive categories. Islands are typically defined in terms of their exclusion from the infrastructures that support modern societies. According to the second edition of the *Oxford English Dictionary*, the word *island* can indicate the act of isolating and insulating or an object or individual that stands out by itself. John Gillis writes that "islanding" is a Western way of understanding a world that appears shapeless and directionless, one that assigns meaning to individual, bounded things and regards the in-between as a void.[3] This holds particularly true for the way we envision geographical islands: isolation and boundedness are the two factors understood to make them special.[4] Though recent studies have documented the interconnective and archipelagic nature of geographical islands, reframing the ocean as a "sea of islands," the solitary island remains the privileged metaphor for isolation in the West.[5] Opposed to this is the network, understood as a system of interrelationality made possible by protocols, the sets of rules and standards that enable exchange.[6] When infrastructural systems fail, we see them as losing their status as networks: a broken route is not a part of the network; it is off-line and no longer networked. If *island* has come to signify disconnection, *network* often simply indicates connectivity.

The network versus island schema encompasses two divergent ways of looking at the world. To see the world in terms of islands is to pay attention to the isolation of individual elements, whereas to see it in terms of networks is to pay attention to that which connects. This obscures significant material dimensions of our communications networks, especially in the postcolonial Pacific—including the connectivities that geographical islands enable, the importance of island topographies for network development, and the imbrications of transoceanic communication with flows of culture and transportation. Viewing the world through this lens also makes it difficult to see how network infrastructure brings only partial, contingent forms of connectivity and needs continual upkeep and repair.

The network versus island schema also has a tangible impact on the discursive and political developments of undersea cable networks. A conceptualization of the island as inherently isolated is used to justify extensions of network infrastructure to islands without cables. When cables are portrayed as automatically connective, they appear to bring with them the full force of global information flow. A recollection of New Zealand's first telegraph proclaims: "Cut off by 13,000 miles of sea, and six months in time, from their homeland, British colonists living in remote New Zealand in 1876 suddenly found themselves linked with England—of nostalgic memory—by submarine cable. Thirty-six years of isolation was at an end."[7] Even though New Zealand had been linked to the homeland in numerous ways—via ships, migrations, news circulation, and the postal service—rhetoric about the telegraph reinforced the perception of the island as "isolated" prior to electric connection, obscuring and displacing other potential modes of connectivity. This view positions cable technology as bringing a new and complete connection, despite the fact that it was used primarily by the elite. Cable connections are sites where publics articulate aspirations to participate in global circulations, but they also stir up fears that the last isolated spaces are being lost, producing a cultural anxiety that islandness will disappear.[8]

This chapter both testifies to the strength of this imagined geography, tracing the ways it has been mobilized across the Pacific, and complicates that geography by charting a broad range of relationships between Pacific Islands and cable infrastructure. As an alternative to the network versus island schema, the chapter reveals how networks and islands are mutually constituted. Networks have been hooked into the cultural geographies of islands and in turn have been shaped by those geographies in important ways. To illustrate the co-constitution of networks and islands, this chapter traverses four geophysical islands, narrating each as a node in an archipelago. The first half of the chapter describes the networking of two of the most significant nodes for transpacific cable systems—Guam and Fiji—which was facilitated in part by the legacy of colonial presence. More signal traffic moves through Guam than through many nations: the island's geophysical insularity, the military's presence on it, and its state of geopolitical exception provide a layer of insulation difficult to achieve on the mainland, and this network geography builds on Guam's history as a point of interconnection for East-West trade. Fiji was also a critical node in the early twentieth century, but in the 1980s two political coups transformed it into a socially turbulent environment. Fiji's struggle to sustain signal traffic highlights a challenge that islands continue to face: they cannot solely link to existing networks as an endpoint for signal traffic, but must fully interconnect with them, cultivating reciprocity between local or regional circulations and global traffic. If Guam's cable history challenges our understanding of islands as isolated, Fiji's challenges the conception of networks as simply connective.

The second part of the chapter turns to an uncabled island and a recently cabled one, charting the ways in which islands aspire to become networked. Triangulation, a form of interconnection that involves positioning oneself as a mediating point for existing circulations, is a key approach to doing so. Yap, a state in the Federated States of Micronesia, was once an important network node, but its technological connectivity now depends on U.S. aid and its ability to triangulate flows between Guam and Palau. Yap's history reveals how tenuous network connections can be without ongoing support: many islands rely on external assistance to meet and maintain the required protocols. The case of Tahiti's first undersea cable, the subject of the final section of the chapter, offers a model for how we might imagine cables not simply as a connection to an isolated location, but as a site where interconnection and reciprocity between different networks can be established. Rather than seeing islands and networks as mutually exclusive, this chapter illustrates how the material development of islands' insulation and interconnection is critical to their inclusion in undersea networks. Islands are sites where transpacific circulations are grounded and protected from interference; they are nodes from which publics can tap into, reroute, and triangulate flows as they pass by. Through these stories, the chapter complicates distinctions between connected and disconnected places and further substantiates the topography established throughout the book: technical networks of cable systems are both anchored in and stabilized by their surrounding environments.

Guam: Networked Island

About four hours by plane from Tokyo, Manila, and Seoul; five hours from Taipei and Hong Kong; and seven hours from Honolulu (direct flights are available from all of these cities), Guam is a point of interconnection between East Asian and American transportation routes. I arrive on the flight from Hawai'i at A. B. Won Pat International Airport on a hot summer afternoon, where I am met by the son of my mother's brother's wife's cousin (everyone seems to have a distant connection to Guam). While guiding me on a tour of Tumon Bay—whose waterfront is lined with towering, extravagant hotels; upscale designer stores; and seedy strip clubs—my relative asks what brings me to the island, since "it's not on the way anywhere." As I describe the cable networks that position Guam as a center for Pacific Internet traffic, I think about how this island, which is home to approximately 180,000 people, is actually on the way to a number of places. It is a temporary destination for many Japanese and Korean tourists, Micronesian islanders looking for an education or a job, Americans stopping for a tour on one of the two military bases, and travelers on a layover between cities. Guam is a meeting point, a site of exchange at which different kinds of flows do not simply terminate but are interconnected.

These interconnections are enabled in part by Guam's insularity and its exceptional political status. Guam is defined as an *insular area*, a U.S. jurisdiction that is part of neither a state nor a federal district. It exists at the country's forefront and its periphery, paradoxically "an island that is within yet without the United States."[9] It is branded as the place "where America's day begins" (the first part of U.S. territory to greet each calendar day) and the "tip of the spear" (a point of American military power).[10] Because Guam is not a U.S. state, the companies and people who interconnect here are not required to adhere to all federal laws—they are insulated from the feedback loop of democracy. All of the people born on Guam, whether Chamorro, Filipino, or Micronesian in heritage, are U.S. citizens. However, they are not allowed to participate fully in American democratic life: they are not allowed to vote in presidential elections, do not have U.S. senators, and have only one nonvoting delegate to the U.S. House of Representatives. If the cable companies were to engage in activities that upset locals, they would have no direct line of appeal to the federal government. In addition, Guam's physical insularity makes it an ideal site for cables: here companies feel that their circuits will be safe from potential interference (they won't have problems gaining concessions or being taxed, as builders fear in the Philippines or Japan). Guam's remote location, its proximity to a huge undersea drop-off (the Marianas Trench), and the relatively little coastal boat traffic naturally protect cables from fishing or anchoring. The systems here do not even need to be buried under the beach. More cables have landed on Guam than at any other American location in the Pacific—including Hawai'i and California, two major hubs for signal exchange—and as of the time of writing the island has more capacity for international signal traffic than either of these states.

In many ways, Guam still appears to be a typical island, seen by its residents in terms of its isolation rather than this interconnectivity. Guamanians have an acute sense of their marginality—they identify themselves as American but do not see themselves represented in the images of Americans that they watch. Guam lies beyond the frame of many maps of the United States. As scholars have observed, the perception of islandness itself is a cultural process that says as much about one's culture as about the geographic space that is being described. Elizabeth DeLoughrey argues that even though we might all think of ourselves as inhabiting islands surrounded by water, we are more likely to perceive the islandness of spaces inscribed in the history of colonial movements: in "the grammar of empire, *remoteness* and *isolation* function as synonyms for island space and were considered vital to successful colonization . . . their remoteness has been greatly exaggerated by transoceanic visitors."[11] The vision of the isolated, fertile island helped naturalize colonial dominance; today seeing Guam as isolated, peripheral, and helpless without U.S. systems of interconnection naturalizes American dominance. Here I offer a competing historical narrative of the island, tracking how Guam came to be integral to Pacific circuits due to its balance between insularity—facilitated by the U.S. military's establishment of the island as a strategic space—and interconnections with traffic through the region. Via these spatial developments, communications power has been embedded in the island's topography, giving traction to transpacific circulations.

Guam became a site of interest to colonial powers due to its location at the nexus of atmospheric and oceanic currents. When Magellan first crossed the Pacific, he missed most of the other islands between South America and Asia: these currents brought him almost directly to Guam in 1521. The Spanish subsequently controlled the island and used it as a coaling point from the 1500s on. Guam became a stopping point in transpacific trade—a location that mediated East-West currents—and was a strategic location for the next five hundred years. After the Commercial Pacific Cable Company laid its first transpacific line in 1903, the island emerged as a site of interconnection, a place where one kind of network could be leveraged to facilitate the development of another. The fact that the cable system was open (not simply limited to the military) meant that other private companies were allowed to interconnect at Guam; this helped to generate additional cable traffic. In 1905 the German-Netherlands Telegraph Company connected Guam to Yap and Shanghai, and in the following year Guam was connected to Japan. The telegraph cable made Guam attractive as a stop for the Pan Am Clipper and the site of a Pan Am Hotel and was seen by the U.S. Navy as having a potential role in supporting the development of Guam's port.[12] The island was not an endpoint for traffic but instead was imagined as a hub: in his annual report of 1915, Guam's governor wrote of its centrality: "By a glance at the map it may be seen that one quarter of the population of the world lies on a rough semicircle of which the meridian of Guam is the diameter, and Guam itself the center."[13]

Guam's emergence as a networked island was made probable by its historical connections not only with oceanic currents and transportation systems but also with American colonial investment, which leveraged a geopolitical imagination and extension of the U.S. Pacific empire.[14] Although the United States refused to build a state-sponsored transpacific cable, the Commercial Pacific Cable's landing was nonetheless enabled by the U.S. acquisition of Guam during the Spanish-American War only five years earlier. The offer to lay the cable "without subsidies or landing licenses [was made] on the grounds that the Pacific Ocean was a 'navigable water of the United States.'"[15] At the inception of American control over Guam, the island was placed directly under the Department of the Navy, which was given absolute authority over it.[16] The land for the cable station was leased from the Naval Station, and in the early days of its operation, the Navy played a large role in supporting the cable station's operations. The 1904 governor's report noted the Navy's "desire to assist the [cable] company in every practicable way in all of its undertakings."[17] In return, it was by the "courtesy" of the superintendent of the cable company that the naval station was informed of the state of war between Japan and Korea as well as the conflict's progress.[18] As the Navy's control over the island tightened, communications development was made more clearly subject to Navy oversight—the Navy censored the cable several times during World War I.[19] Subsequently, the governor of Guam established a firm definition of the cable company's subservience; he determined that the military character of the island allowed the United States to take control of any part of the cable business at any time. All persons connected with the cable station would have to be U.S. citizens so that they would not develop conflicting interests during time of war, and the government would find suitable employees if the company could not.[20] The cable was used to facilitate important financial transactions for the military, and later the cable company served as a defense contractor.[21] The Navy's presence was key to the maintenance of cable traffic: it provided a blanket of security shielding the traffic from interference—at least until World War II, when the cable station (along with military installations) was bombed.

In the postwar era Guam became an unincorporated territory of the United States, gained a civilian government, and began to fully develop as an infrastructure hub: it was a stop for two major airlines and two major shipping lanes.[22] The Navy invited RCA Communications to set up commercial radio-telegraph and telephone service for the island. After a little over a decade without any undersea cables, Trans-Pacific Cable 1 (TPC-1) was extended between the mainland United States, the Philippines, and Japan, establishing Guam as a critical intra-Asia switching point. Shortly after, the South-East Asia Commonwealth Cable (SEACOM) system linked with TPC-1, bringing Singapore and Sydney into Guam's immediate network and solidifying it as a point that interconnected north-south and east-west transpacific lines. These were followed by TPC-2 in 1974 and a cable to Taiwan, TAIGU in 1981, both of which strengthened Guam's intra-Asian connections. The operational costs of these systems were lessened by the existence of water, power, and transportation infrastructures. They were privately operated yet, like the telegraph network, benefited the military, which continued to provide both a layer of insulation and a reliable source of traffic. In turn, the cable system generated a gravity that drew in other flows and was used to call for further infrastructural concentration, including the development of better air transport networks.

Although many islands that were networked early were later passed over, Guam continued to give traction to both north-south and east-west signal traffic, in part because these currents could connect with each other on the island (map 5.1). In the fiber-optic era, Guam's cable infrastructure advanced significantly: TPC-3 in 1987, China-US in 2001, and Asia-America Gateway in 2009 were new U.S.-Asia networks; Guam–Philippines–Taiwan (GPT) in 1990 and Guam–Philippines (GP) in 1999 were new intra-Asian networks; and PacRim West in 1995, the Australia-Japan Cable in 2001, and PPC-1 in 2009 were new Australia-Asia networks. Guam was their meeting point. In 2010 the military laid HANTRU-1 from Guam to the Marshall Islands to facilitate remote defense: though the majority of traffic on Guam's undersea cables is not defense related, the interconnection of military operations continues to depend on private companies. If the island's networks were disconnected today, it would disrupt not only military operations and transpacific Internet traffic but also the operation of the island's port, the flights that land on Guam on their way from Japan to Micronesia, and the weather reporting for much of the region. The perception of security nonetheless continues to insulate Guam's networks, leading companies to believe that their traffic will be safer there since the U.S. government continues to retain oversight (and is perceived as less liable to tax cable traffic than other governments). The military presence has contributed to a unique organization of power: power is enacted not through the hiding of infrastructures (as in Hawai'i), the manipulation of bureaucracy (as in California), or the construction of walls twelve feet high (as in the Philippines), but through the visual display of American military might and an indirect exercise of control.

In Guam's history, oceanic circulations have been overlaid by subsequent

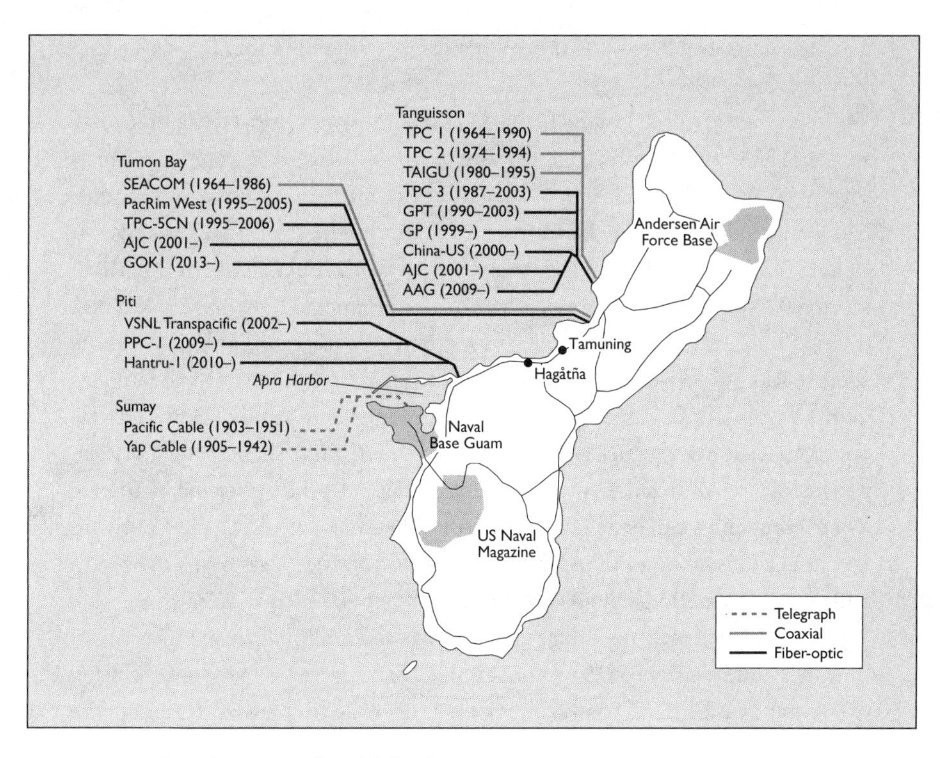

MAP 5.1. Guam's transpacific cable landings.

cultural circulations, and over time, infrastructures were built up in the area to channel, reroute, and direct all of these movements. Guam's success is due in part to its ability to interconnect military and private interests. Systems owned and operated by the military that remained completely insular, such as Guam's early Navy radio stations and the Johnston Atoll cable system, did not develop as hubs for commercial traffic. Guam's success was also driven by private companies' desire to interconnect with one another and to establish the island as a place where this could safely occur. During the first half of the twentieth century, the ability to do this was limited, as the U.S. Navy controlled the island and prevented extensive external development, yet Guam nonetheless became a hub for telegraph traffic. The difficulties encountered by Pacific islands that switched almost entirely to commercial traffic, such as Fanning Island and Fiji, nonetheless reveal the significance of defense in justifying and sustaining cable systems. Guam's continued success at interconnection means that even though the island has some of the best international capacity of the United States, it has not needed to develop many local sources of feeder traffic (that is, signals that originate on Guam and help sustain the island's networks). As one cable manager describes it, from Guam all his company sees is a set of pipes leading outward.[23]

Guam is a networked island—an island that remains critical to the transoceanic movement of information. As signals move between Europe, the Americas, Asia, Australia, and Africa, they intersect numerous networked islands: Fiji, New Zealand, and O'ahu in the Pacific; Bermuda, the Canary Islands, St. Helena, and Puerto Rico in the Atlantic; and Sicily and Cyprus in the Mediterranean. The stereotypical imagination of an island as a remote tropical destination obscures this centrality. This history of spatial manipulation, via the coordination of private investment in interconnections and strategic interests in establishing redundant communication, sculpts contours in the island's topography that attract further interconnections. Guam has become a critical node: a place that is not just the sum of individual traffic routes but also a site where companies can benefit from existing structures; where blocked outward transmissions can be rerouted (although inward traffic is less easily diverted); and where economic surpluses are more easily generated. Its power extends beyond the grasp of any individual force such as the military, private investment in trade routes, and the oceanic and atmospheric currents. This concentration of communications resources makes the island of Guam a pressure point in the cable network, where—as at many cable landings—local actions and environmental forces can produce disproportionate effects on its operation.

Fiji: Regional Hub

Guam has been the pressure point for the North Pacific; Fiji has long occupied that role in the South Pacific. On both islands, colonization set the groundwork for cable development. In 1902 Fiji preceded Guam as one of the first Pacific islands to be linked with telegraph systems via the British All-Red Line. These cables were well supported by existing infrastructure, including buildings, ports, and communities (this was not true, for example, on Fanning Island, Midway Island, or Yap). Fiji's networks and staff were actually better supported than those on Guam: although early cable workers on Fiji found a home in the colonial settlement, the Commercial Pacific Station staff struggled to harness Guam's resources, even with assistance from the U.S. Navy. Today, the situation has been reversed. Fiji has gained independence and currently has two large-scale transoceanic cables (each of which is capable of carrying 2.7 terabytes per second) that it is struggling to connect with local and regional circuits. Guam has twelve links extending from it, which cumulatively have a capacity of approximately 21 terabytes a second—enough to carry simultaneous phone calls from everyone in the United States.[24] This is almost four times as much capacity, despite the fact that Fiji's population is almost five times greater than that of Guam.

As was the case on Guam, Fiji's cable landing facilitated the accumulation of local connective resources, establishing a wired topography in which the island's position as a stopping point was leveraged to build social and economic connections with larger countries (which in effect subsidized Fiji's communications costs). But unlike those on Guam, Fiji's telegraph cables were used extensively to gain investment for island development, primarily in Fiji's agricultural sector and in its regional telecommunications. The emergence of wireless telegraphy introduced an infrastructural geography to Fiji that was absent in Guam's early period: Fiji became a regional hub for numerous other Pacific islands. Due to the minimal investment needed for wireless, Fiji was an ideal interconnection point for islands on which a cable to Australia or Canada would have been economically unproductive. Local resources for support were developed, including a wireless telegraph training school in Levuka, Fiji's first capital.[25] Images and narratives from the 1940s suggest that the role of Pacific islanders in Fiji's telecommunications network prior to and during World War II was significant: this school included not only Indian and Fijian students, but Gilbertese, Ellice Islanders, and Solomon Islanders. When high-frequency radio emerged, it reinforced these local feeder connections, providing coast watch services and telecommunications to other South Pacific islands.

By the 1950s and 1960s Fiji had become a true Pacific communications hub.[26] In place of British colonial or military support were the corporate interests of Cable & Wireless, which used the island as its Pacific switching point and secured the development of Fiji's wireless geography. This geography was dependent on and reinforced the cable connections: wireless provided traffic for the cable network and motivated further investment in cable infrastructure (figure 5.1). In 1962 the Commonwealth Pacific Cable (COMPAC) was initiated, again solidifying connections between Fiji and the larger economies of Australia, Canada, and New Zealand. Although initially the cable was intended to stop at two other islands on the original telegraph route, these were eliminated as landing sites—transpacific telephone cables had begun to hop over intermediary islands. Fiji's place in the COMPAC system was assured both due to the presence of wireless feeder links and its alignment with Cable & Wireless's network. Cable ships and a maintenance depot supporting the entire South Pacific region were located at Fiji (with a largely Fijian crew, the depot boosted the local economy).

In the late 1960s and after Fiji gained independence in 1970, the govern-

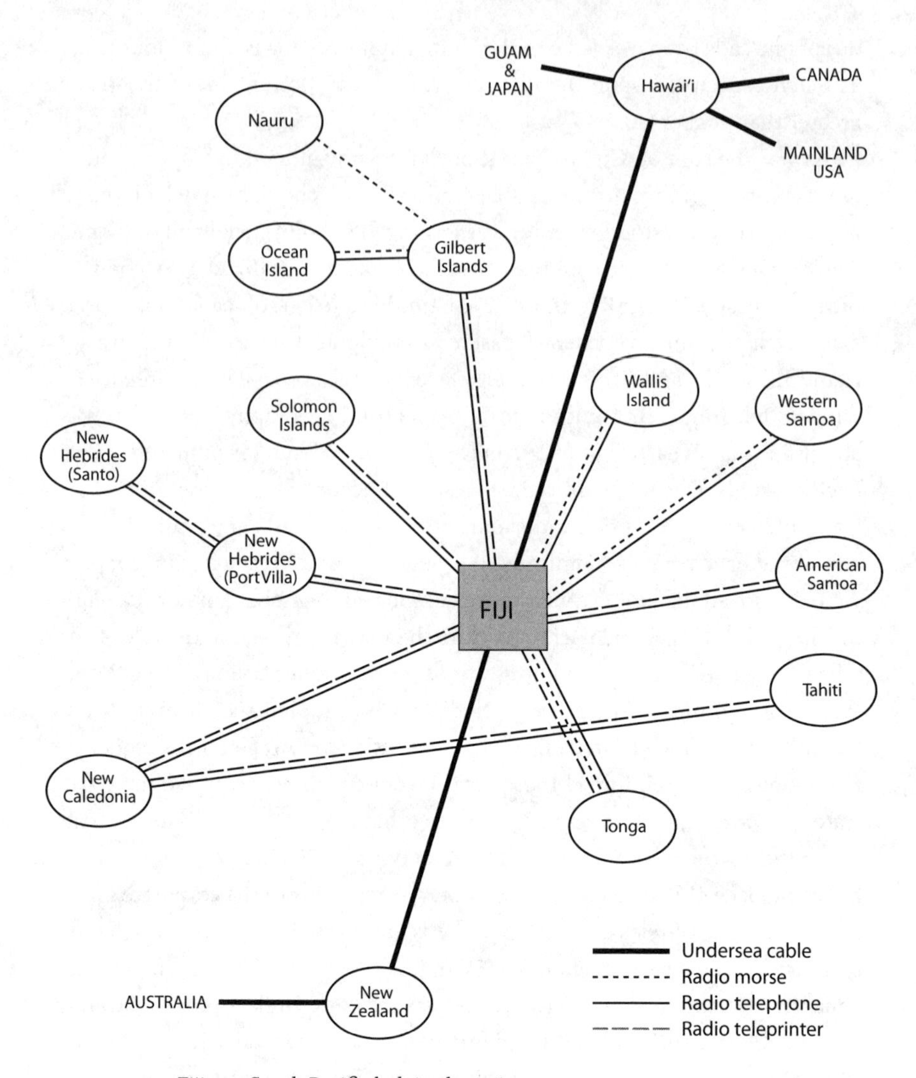

FIGURE 5.1. Fiji as a South Pacific hub in the 1960s.

ment increasingly took over Cable & Wireless's operations and used wireless to further strengthen Fiji's role as a site of regional interconnection in the South Pacific. As on Guam, satellite links supplemented but did not replace wired connections. In the 1970s the Pan-Pacific Education and Communication Experiments by Satellite (PEACESAT) program established Fiji—in particular, the University of the South Pacific—as an educational satellite hub for other Pacific islands. Fiji also established a Telecommunications Training School that prepared several cohorts of communications professionals from around the Pacific. These links, along with vested interests of Cable & Wireless, made it financially viable to include Fiji in the 1984 transpacific cable system, ANZCAN. Like earlier cables, ANZCAN provided secure links to developed countries, and Fiji continued to mediate the access of other Pacific Islands.

In the first seventy years of its telecommunications development, Fiji emerged as a hub for regional islands in the South Pacific via the connection of its cabled geography (wired links that facilitated exchanges between Fiji and more developed countries) with its wireless geography (which leveraged these transnational circulations to situate the country as a hub) and the development of local training to support the industry both locally and regionally. In the South Pacific, Fiji was the only site that had redundant routes both to the north and to the south and that was able to serve as a regional transit point, triangulating transpacific and regional circulations. Not only did Fiji depend on these signal flows, but—as was the case with Guam—major Western powers also depended on the smooth operation of the island. In contrast to Guam, however, on Fiji the perceived stability of the island was ensured not by military presence, but by the corporate investment of Cable & Wireless. As a critical intermediary node, these forms of insulation and interconnection were important: local infrastructure, including local workers, needed to function flawlessly for global traffic to run.

After Fiji gained independence, the telecommunications companies were aware of the country's looming political tensions: it no longer appeared to be a safe location, where the companies' signal traffic would be insulated from social conflict. An engineer involved in ANZCAN remembers that "there was a lot of concern about the unstable political environment in Fiji," but the company had to limit the lengths of undersea cable segments, and there weren't many other options, unless they chose a "remote island" at which there would likely be maintenance difficulties.[27] Fiji's two political coups in the late 1980s dramatically shifted the communications environment. During one coup, the military took over the cable station and blocked all outgoing traffic. Though the traffic was eventually restored, the instability had a lasting economic effect. Prior to the coups, there were discussions about how a broadcast training program might build on the island's successful telecommunications network. One plan situated Fiji as a hub for television, which—modeled after the telecommunications industry—would route service to the other islands and better position the country as a site for regional and transoceanic interconnection.[28] Although negotiations for a television station had been forthcoming, partner companies left the country after the coups. At the moment when the government briefly considered tapping into the island's central position in the cable network to make it a hub for media dissemination, the political instability and the withdrawal of international investment thwarted these plans.

Much of the 1990s were spent trying to regain stability and draw in the economic circulations that had left the country. Many Indo-Fijian employees of Fiji International Telecommunications Limited (FINTEL), which was partly owned by Cable & Wireless, had left the company. Fiji was able to do relatively well in maintaining its wireless network, which needed low levels of financial investment and did not require extensive international backing. A demand assigned multiple access satellite network was set up in 1993 to give Fiji direct satellite routes to numerous Pacific island nations as well as to Canada, the United States, Australia, New Zealand, Hong Kong, Singapore, Japan, and Korea.[29] However, this was operated by Telstra, an Australian company. During this period, interconnection at Fiji was based almost entirely on wireless links, and the island lost its place in the cabled geography of the Pacific. In the early 1990s, when the PacRim cable was designed, FINTEL considered hooking up to it on its way across the ocean, but the cost was prohibitive since Fiji had not recovered from the postcoup loss of transnational investment.[30] In addition, the cable companies viewed the island as dangerous and not adequately insulated to safely transmit international traffic. As a result, PacRim curved around Fiji in a U-shape: it was the first transpacific cable between the Americas and Australia and New Zealand that did not use Fiji as an intermediate link. Fiji was put in the precarious position of having invested almost all of its money in developing as a wireless hub, at a point when traffic was switching to fiber-optic cable and satellites were becoming increasingly difficult to use for the Internet. Without a high-speed cable, Fiji now had to bear the full cost of transmission to major economies, which lessened the country's advantage over Australia and New Zealand as a point to triangulate regional circulations. The loss of PacRim was a moment of crisis in Fiji's telecommunications history, one that might have positioned the country permanently alongside other Pacific islands with lower quality communications.

In the late 1990s, when the Southern Cross Cable Network (SCCN) was being designed, the Fijian government and FINTEL made a decision to reposition the island at the center of a cable network. This was done without the help of a colonial or corporate support system, but the decision to land in Fiji was nonetheless aided by the continuing mediation of Cable & Wireless and the existence of landing facilities. One cable route engineer reported his skepticism about the country's stability: "We were in two minds as to whether the main cable should be landed in Fiji. I guess the answer is really dictated by economics because it was much cheaper to have a full landing."[31] Fiji's decision to

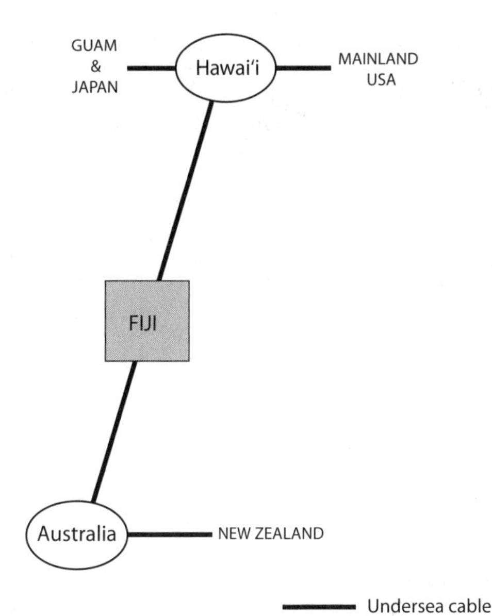

FIGURE 5.2. Fiji abandoned its wireless connections in 2001 and linked to other countries solely by undersea cables (shown here).

connect to Southern Cross involved generating an enormous amount of capital, $48.5 million, for the project. The chairman of FINTEL at the time wrote: "This investment makes Fiji the only small island country in the South Pacific to make available such high-speed bandwidth to its customers and opens up a huge potential to the development of the entire ITC [information and communications technology] industry in Fiji. . . . This potential would not have been possible without the SCCN. . . . I also wish to stress that the investment in the SCCN was a heavy financial commitment by a relatively small company like FINTEL."[32] At the same time as FINTEL and Fiji took on the financial commitment of the cable, they also had to sacrifice their less economically productive regional satellite links, withdrawing completely from the wireless geography and their role as a Pacific hub (figure 5.2).

The huge debt acquired from the cable, the sudden lack of any kind of interconnecting traffic, and the costs of continual maintenance led to a push in the late 1990s and early 2000s to ground the network in local circulations and develop new sources for signal traffic.[33] One of the initiatives of this period was "Bulawood: The Hollywood of the South Seas," a plan to replace the lost telecommunications links with signals generated by media production. The Fiji Audio Visual Commission (FAVC) was created to stimulate the development of hybrid forms of information and communication technologies (ICTs) with visual media. The *audiovisual* was reconfigured to include call centers, back-office operations, and multimedia services in the FAVC's purview.[34] The

FAVC imagined that studio cities would be built, where ICT and film production companies could be housed in close proximity and interconnected. Fiji's government also created tax-free zones where ICT companies could develop free of income tax for ten years, a strategy of insulation designed to draw in economic circulations. Efforts were made to establish call centers and other outsourcing jobs in digital media.[35] The design of studio cities, tax-free zones, and ICT parks—all circumscribed areas that could facilitate interconnections between local and global circulations—were attempts to leverage the infrastructural formations of the colonial era and to develop technology and media industries as a site of production for data traffic. Although the lack of economic and political stability in the postcoup period initially delayed the internationalization of media production, ultimately it laid the groundwork for revisioning communication infrastructures as sites for transnational technology and media production.

At the same time, Fiji had to develop visual and rhetorical frames to make undersea cable infrastructure public, and the country had to portray itself as a critical hub rather than as being isolated from global networks. The manager of industrial zones at the Fiji Trade and Investment Bureau told me that "Fiji is the hub of the Pacific, in terms of shipping, in terms of airlines, and in terms of communications," and that making these connections apparent to potential investors is necessary to the country's development.[36] During an interview, Florence Swamy, then acting CEO of the FAVC, reiterated that since the media industry is based on information technology, letting potential investors know about Fiji's interconnections is a crucial part of her job.[37] Indeed, the vision statement of the FAVC, titled "Fiji: The Audio Visual Hub of the South Seas," frames the island in terms of its interconnectedness and centrality to transpacific flows. This rhetorical positioning, along with the FAVC's plan to develop an audiovisual training school and other regional film resources (based on the telecommunications industry model), represents an attempt to construct Fiji not simply as a destination, but as a place where information flows can be interconnected in the South Pacific—like Guam. In many places, the perception of connectivity, and by extension the perception of cable infrastructure, is as important as actually obtaining a cable.[38]

Despite the need for connectivity, media producers and transnational investors are still drawn to the country by its image as the quintessential remote island, which is how Fiji has been presented since the colonial era.[39] The majority of the films produced on the island—from *The Blue Lagoon* (Frank Launder, United Kingdom, 1949) to *Cast Away* (Robert Zemeckis, United States, 2000)—emphasize Fiji's islandness: its disconnectedness not only from communications systems but also from Western culture and practice. Those who seek to draw existing flows across the Pacific to Fiji must both mobilize a set of imaginations that play to Western fantasies of disconnection, which will lure more tourists and investors to the islands, and assure their users that Fiji is also a networked island linked to the rest of the world. During an interview, the CEO of Tourism Fiji told me that this is precisely the problem the island has to confront: investors, tourists, and media producers all want to be as connected there as in any other location, yet at the same time they want to be able to feel selectively disconnected.[40] For those working in Fiji's media and technology industries, participating in the global economy is not just about getting a cable connection. It is about mobilizing historical developments in connectivity alongside an imagination of islandness (two developments that were facilitated by colonial interests in the region) to accumulate resources in Fiji.

Like Guam, Fiji initially depended on a colonial presence to help develop and insulate the routing of international signal traffic, but this eventually changed as British support was withdrawn from the island. For many years, this historical communications topography helped Fiji maintain its centrality as a regional hub and site of wireless interconnection. Although Guam's success depended on interconnecting U.S.-Asian, intra-Asian, and Australian-Asian circuits, Fiji became a switch between the Pacific islands and cables between Australia and the United States and Canada, an interconnection point that linked regional and transpacific traffic and that triangulated other islands with the mainland. However, when the colonial presence of the British and the corporate presence of Cable & Wireless receded, the island lost a layer of insulation that had made it appear safe for communications signals. To maintain and fund its contemporary cable connection, Fiji had to become more than simply a reception point, and the country began to look for sources of signal traffic that would feed into the transoceanic network. Today they have also begun to triangulate regional communications once again: undersea cables have been laid to both Tonga and Vanuatu. This project is one not simply of connecting, as Fiji already has cable links to Australia, New Zealand, and the United States, but of interconnecting: leveraging domestic and regional circulations to maintain the island's place in the undersea network.

Yap: After Cables

My arrival on Yap is an immediate contrast to my arrivals on Guam and Fiji and quickly shifts my attention to the ways that Yap has diverged from the other two islands since the three were hubs in the early twentieth century. When I step off the plane, the first thing in my line of sight, and the largest structure on the runway, is a bright yellow truck with reflective lettering illuminated by airport lights: "Yap International Airport Crash & Rescue." It is three o'clock in the morning, and I am disembarking from one of only three weekly flights to the island. Although major airports around the world are typically closed at this hour in local time, Yap's flights arrive and depart between 1:00 and 4:00 AM so that passengers will be able to reach their continental destinations during the daylight hours: the periphery of the network is open only during marginal times of operation. Guam's international airport has over seventy arrivals and departures a day (triangulating Asian traffic), and Fiji's Nadi airport (a connection point for Australia and New Zealand's traffic) has around fifty, but flights arrive in Yap only two days a week. It does not interconnect circulations between mainlands or circulations between continents and islands. Rather, it triangulates interisland currents and is a stopping point between Guam and Palau, a more popular destination for divers.

Andrew is here to pick us up in a 1980s van to drive us to the hotel where he works. Among those arriving with me are a bird-watcher from the United Kingdom who is hoping to start a regional birding tour, a software developer on vacation from London, and a team of German scuba divers here to see the island's famed manta rays. Everyone is stopping on Yap as part of a vacation to other Micronesian islands; they have been attracted, like the tourists in Fiji, by an image of the island's isolation, a perception heightened by the fact that we are all stuck here for the three days until the next plane comes. The relative smoothness of our visit, however, has been made possible by Yap's continued connections to external funding. The roads from the airport to town are well paved and, like the crash and rescue equipment, have been paid for by U.S. tax dollars (between 2004 and 2009, the Federal Aviation Administration gave over $30 million in grants to Yap for airport improvements).[41] Although support has been made available for new infrastructure projects, the remnants of infrastructure failure remain. Wrecks are scattered about Yap's islands. Many are unexcavated plane crashes from World War II. The ruins of a Continental Airlines/Air Micronesia Boeing jet from 1980 remain in the jungle adjacent to an old runway. Small planes flying between the islands are the most susceptible to environmental fluctuations. If conditions are not exactly right on Yap, sometimes flights will overshoot the island and continue onward to the more important destinations: interisland circulations are the least protected. The heightened vulnerability of such systems—given the lack of insulation and limited amount of traffic, which generates less profit per visitor—means that Yap, with a population just over 11,000, must go further to meet the standards for interconnection with existing communication and transportation networks, often relying on international aid.

A century ago, when the first transpacific cables were being laid, Yap was a Pacific communications hub, with a position in the network equivalent to Guam and Fiji. After colonization by the Germans in the early 1900s, it was positioned as the junction point of three cables linking Guam, Shanghai, and Manado (in Indonesia). By 1914 the Germans had strung 18,000 miles of cable, connecting Yap, like Guam and Fiji, with a vast colonial network.[42] From Guam, the United States relied on the German-Dutch network to connect to Shanghai—the network formed an important bridge between America and China. As the historian David Geddes describes the situation, since "a cable base was a prime instrument of economic, military, and political power . . . the USA never looked kindly on the German ownership of Yap."[43] During World War I the Japanese seized the island, and Yap became not only a "cable threat" but a threat to the sea route between the United States and the Philippines, another important line of connection.[44] Between 1920 and 1921, the possession of Yap, its cables, and its coaling station were the subject of a dispute between the United States and Japan and involved in a host of international issues. Japan claimed responsibility for the island under a mandate from the League of Nations. The United States sought to put Yap and its cable station under international supervision, rather than under the control of a single nation. After a year of deadlock, the United States and Japan came to an agreement that split the network: the Yap-Guam cable went to the United States, the Yap-Manado link went to the Netherlands, and the Yap-Shanghai link went to Japan.

Had historical forces conspired otherwise, Yap might have taken the path that Guam did, becoming a point insulated by military presence and a point of interconnection between Japanese and U.S. networks. Or it might have ended up like Fiji, a regional transit point triangulating continental powers and Pacific islands. None of these nations, however, built up the necessary infrastructure to interconnect its cable station. The cable companies ceased to operate their lines to Yap, as they did with Fanning Island, Norfolk Island, and Midway Island, focusing instead on other locales. Yap was skipped over as future telegraph, telephone, and Internet cables extended across the Pacific, and because the island did not continue to host such circulations, the pull of its communications topography disappeared. From the 1920s onward, Yap continued to rely on wireless technologies, connecting through network hubs to reach the rest of the world.

As satellites become a secondary technology for Internet transmission, islands that were dependent on this infrastructure, such as Fiji in the 1990s and

Yap today, are put at an increased disadvantage. Although in the 1970s, satellites were seen as an ideal and appropriate technology for island nations, and a possible facilitator of socioeconomic development, today satellites are no longer cost-effective, and Yap cannot afford to connect to a high-speed Internet infrastructure. Indeed, the island's telecommunications company could purchase more satellite bandwidth, but Yap does not even use the low amount of capacity already available: there are no local or regional economic circulations to justify it.

In the absence of such infrastructures and interconnections, the basic expense of running networks on islands such as Yap are much higher than they would be in urbanized or developed locations. When I take a cab from the capital of Colonia to visit Yap's satellite station, I find that it is what looks like a domestic residence with almost no security. As is the case on other Pacific islands, on Yap communications companies must overcome maintenance challenges to insulate signal traffic, including managing temperature, moisture, and the level of salt in the environment. They also must organize training and technical upgrades, which is difficult given an exchange rate and distribution costs that make equipment more expensive. Reliable energy infrastructure can also be a problem, especially given the need for power-intensive air-conditioning. One cable entrepreneur described to me the difficulties he encountered in planning a system for Tonga: since there would be not enough storage on Tonga to get fuel directly from Singapore, the country would have to import it from Fiji, which in turn would get it from Singapore, increasing both time and expense for Tonga.[45] Infrastructural limitations cascade for those on the outer edges of the network, and developing islands simply have to do more groundwork to interconnect.

Despite the fact that Yap doesn't currently need the capacity, and despite the limitations that make it difficult for the island to maintain the connections it has, there are still calls for a cable network. In 2011 there were talks between Palau and Yap about a potential cable system that would link Yap to Guam, following the same route as the planes. Project managers across the Pacific have told me that cables are important, even essential, to the islands: cables can result in increased reliability for weather reporting, including tsunami warning systems, and will support higher bandwidth applications for less money, assisting in development.[46] In 2010, when the HANTRU-1 system was laid, an arm branched off to Pohnpei, another one of the Federated States of Micronesia. Reflecting on the installation, Tony Muller, president of the Marshall Islands Telecommunications Authority, stated: "Connecting to the global fiber optic network promises to be nothing short of transformational for an isolated country such as the Marshall Islands. . . . The HANTRU-1 extension . . . is essential to our nation's development."[47] On many islands I visit, I hear rumors that a cable is coming soon, and the telecommunications companies of many countries call for such developments.

It is not clear where the feeder traffic—whether domestic or regional—will come from to support these cables, and the projects in the meantime would have to be subsidized by governments and external organizations, triangulating existing sets of interests. HANTRU-1, for example, is a link built for the U.S. military between the Marshall Islands and Guam, and its Pohnpeian extension was financed by the U.S. Department of Agriculture's Rural Utilities Service Telecommunications Loan Program. The slow timetable of international granting organizations often does not accord with private cable systems' tight deadlines, and the work of developing a local connection often falls to islanders who have less experience in the process. As one transpacific project manager told the representatives of a local island telecommunications company: "We're coming past. We're building a spur. You guys have to do the environmentals. . . . You have to do the landing permissions. You have to do the permits. We are not involved. Is that clear? Because we can't afford that time and that distraction. It's your country. If you want it, you organize it."[48] These developments can be hard to justify in light of other needs, and as Brett O'Riley observes, connecting to islands on the way across the ocean simply "comes down to the good will of particular individuals."[49] O'Riley tells me that convincing Pacific leaders "to devote dollars into an international fiber-optic cable is something quite difficult to contemplate," as they have often less exposure to technology and face enormous challenges in securing adequate food, water, and shelter.[50] After all, even politicians in developed countries are skeptical about funding cable systems. For some islands, given the expense in upkeep, it makes more sense to charge companies for traversing their oceanic territory than it does to establish the layers of insulation required to shelter a cable station and set up domestic infrastructure to support it. One engineer recalls that, in the case of one island, making a payment of "half a million or something" to the locals kept both them and the cable company happy, since the company otherwise might have had to invest a million dollars to divert around the island.[51]

Another challenge confronts island nations such as Yap if they attempt to obtain a cable: government telecommunications monopolies. The government monopoly in Yap, the Federated States of Micronesia Telecommunications Corporation, carries all traffic in and out of the country, which means that the corporation can dictate the price of capacity (which remains high given the significant amount of money required for investments), and this further limits access. Since the 1990s many countries have deregulated international communications, opening up places such as Fiji and Guam to more competition. In some places, such as Singapore and South Africa, cable stations have been deemed "essential facilities," a doctrine derived from U.S. antitrust law that enables the government to regulate critical infrastructures that are not easily replicable.[52] Opening up cables to competition and forcing companies to share facilities has lowered the prices for international signals in many places. In Yap the monopoly remains, and the price for bandwidth, along with the price of a computer, remains out of reach of most Yapese. Even if Yap receives international financing, breaks up its monopoly, brings down prices, and distributes affordable (or free) end-use technologies, the population is dispersed across many islands and will still require domestic wireless technologies. For islands across the Pacific, satellite and radio will continue to be more viable alternatives, especially in the absence of aid funding, creating a two-tiered distribution system.

Networks have altered the topography of the Pacific. Indeed, Yap first became visible to colonial powers because of its potential to be networked. However, Yap's history shows that despite claims that everyone will eventually have high-speed Internet, there are fundamental limitations of maintenance and upkeep that mean not every island will be included in such infrastructure systems. This becomes a problem when the functioning of our social sphere depends on high-bandwidth Internet applications that are accessible via only one of the tiers. Although the push to mobilize cutting-edge systems drives investment in cable infrastructure, a movement toward higher resolution and more interactive content ultimately widens the gap between cabled and uncabled places. Many people in the telecommunications industry suggest that the adoption of new technologies—such as cloud computing, which entails storing one's data remotely (often outside of one's national boundaries)—offer an opportunity for developing countries, whose citizens could then access a wider array of programs and lessen their dependence on domestic infrastructures. Yet, at the same time, this move would make them more dependent on external support to finance cables.

For Yap and the other Federated States of Micronesia, the best chance of becoming an equal node in the network is not likely to entail triangulating flows between mainlands (as it is on Guam) or between transoceanic and island networks (as it is in Fiji), but to involve strategically mediating—and connecting with—interisland flows. However, the small amount of traffic moving in and out, whether via air or sea, will never justify the expenditures needed to maintain multimillion-dollar networks. Without support in developing insulation for its networks, whether via air-conditioning or aircraft rescue, a single node can make the system vulnerable. As a result, cabling many islands in the Pacific will likely continue to depend on the work of governments. In this context, the call, or grapple, for a cable is an attempt to reach out, pull in, and harness transiting currents, and to nonetheless position oneself between existing nodes. It reflects a desire, currently being sounded around the world, to channel some of the smooth flow of power and to ground it locally, even in the absence of local infrastructure or economic circulations that could sustain it.

Tahiti: From Connection to Interconnection

Surf-bound, lonely islet, / Set in a summer sea, / Work of a tiny insect / A lesson I learn from Thee— / For to your foam-white shores / The deep sea cables come; / Through slippery ooze, by feathery palm / Flies by the busy hum / Of Nations linked together, / The young with the older lands, / A moment's space, and the Northern tale / Is placed in Southern hands.
—Ernest Shackleton, "Fanning Island"

In islands' calls for the development of undersea cables, the network versus island schema looms large: islands often use the rhetoric of isolation to justify the immense investment needed for a cable link to the outside world. The cases of Guam, Fiji, and Yap reveal that this initial connection is only part of the story. The critical part of maintaining one's network over time is in interconnecting such technical systems with existing circulations, whether those are other transpacific networks, regional circuits, or local infrastructures. Turning to the story of Tahiti's first undersea cable, Honotua, we see how the recognition of the importance of interconnection might generate a new mode of imagining undersea networks, and perhaps even new approaches to designing them.

For much of its history, French Polynesia, like Yap, was comprised of uncabled islands. Until recently Tahiti relied on wireless means of communication, but due to the scarcity of satellites in the South Pacific, the cost for bandwidth remained incredibly high for its approximately 270,000 residents. In 2010 the Honotua cable—stretching from Tahiti to Hawai'i—was made possible by funding from the French government, granted as part of a wider measure to extend broadband to French citizens and improve the country's digital economy. As was the case with Yap's airport improvements, this tie with a former colonial power and current public funding initiatives subsidized international links. One telecommunications executive claims that Honotua could never have gotten off the ground without government support, since "it's just uneconomic."[53] As on Fiji, in Tahiti the cable is now the primary route for all international traffic, enmeshing the islands in a geographic and technological structure of dependency on U.S. information networks.

FIGURE 5.3. Honotua brings together traditional boats and cable ships. Pacific Network TV.

Instead of framing the cable as simply a technical bridge to a global network, through its development and initiation Honotua was portrayed as a significant cultural interconnection between Tahiti and Hawai'i. A series of videos on Pacific Network TV and Big Island TV News covered the ceremonies of the cable landing, staged in part by the telecommunications companies, at both ends. The videos feature men and women dancing in traditional native attire, with the cable ship in the background (figure 5.3). In Hawai'i, one scene depicts Hawaiians chanting the genealogies of their ancestors who came from Tahiti, while beating on a drum brought over by these ancestors. The mayor of Hawai'i Island then speaks about the cable's revitalization of genealogical connections and claims that it will help Hawaiians understand where they came from. A Hawaiian state senator tells viewers that the cable landing is "like a family reunion." A telecommunications representative describes it as not just a business venture but a "cultural venture."[54] These discourses construct the cable landing as a significant historical moment in which Tahitians and Hawaiians will be drawn together, regardless of who is actually sending signals (and even though many messages will likely be directed back to France).

The cable was also seen as part of, and emanating from, local language and mythology. In Tahitian *hono* means a link and *tua* means the back, backbone, or far ocean. The cable's name reiterates its function: a link from Tahiti across the ocean.[55] During my visit to Tahiti, a member of a technical team tells me about two further symbolic coincidences that reveal how connected the cable is with the local culture. He describes the legend of the Tahitian queen who long ago lived at the cable landing at Papenoo and had traveled to the landing site at Hawai'i: the cable follows her direct path. I ask him if the cable builders knew this while planning the route, and he responds that this historical layering was not intentional but a mere coincidence. He then describes a second coincidence: after the cable had been named, the local mayor informed them that there was a man who lived near the landing with a relative named Honotua, a second fortuitous connection of cable to genealogy. Although Honotua is a business venture that Western tourists will undoubtedly use to connect to their homes from Tahitian hotel rooms, it is locally articulated as having the capacity to transform the relationships between islands. In this way, Honotua offers a model for understanding all cable connections not simply as bridges to a mainland but as interisland networks and sites of reciprocity. Epeli Hau'ofa describes the central and ancient practices of reciprocity, which are "the core of all Oceanic cultures": "For everything homelands relatives receive they reciprocate with goods they themselves produce. . . . This is not dependence but interdependence, which is purportedly the essence of the global system."[56] Linking the system through local customs and regional connections, a cable appears not simply as a connection—a closure of a previously empty gap—but as an interconnection, a site where exchange is established between systems.

As is true on most cabled islands, in Tahiti not everyone is able to connect with the network. Honotua links only the five most populous and most economically "interesting" islands; some segments of the population will remain off the network.[57] The cable itself does not solve the digital divide but must be connected with many other infrastructures, domestic and regional, to provide embodied network experiences. At the same time, the nation now has far more bandwidth than it needs: "We don't need all the capacity," one of the project managers tells me, "we can share."[58] As on Fiji, there are not enough local circulations or feeder traffic in Tahiti to fill up the system. Yet, even though the country has just received a cable, Tahitians are already discussing the possibility of a second. Thierry Hars, of Tahiti's Office des Postes et Télécommunications, tells me in an interview that if the cable goes out, Tahiti will be able to back up only about 5–10 percent of the capacity via satellite.[59] With only one link, it is difficult to develop industries such as call centers that would cease to operate if the one cable went out. Hars hopes that Tahiti will become a hub. "You can see on the map, Hawai'i is a very big hub in the North Pacific," he tells me. "Tahiti could be a hub in the South Pacific." The country could link to Samoa to the north or Chile to the east, triangulating signal traffic to South America. Imagining Tahiti as a hub parallels the dreams of interconnectivity in Fiji and Guam. It is a conception of island as node—as a possible triangulation

FIGURE 5.4. The logo of the South Pacific Island Network. Courtesy of Rémi Galasso, SPIN Network.

point for transpacific hops, a site where regional movements can be channeled to support the network.

Honotua was initially part of a larger cable system, one that took the imagination of island as node to its extreme. The South Pacific Island Network (SPIN) was a transpacific system intended to connect signal exchanges from North America and Hawai'i to those from Australia and New Zealand, using American Samoa, Samoa, Wallis and Futuna, Fiji, New Caledonia, and Norfolk Island as intermediary nodes, which would give redundant links to many of the unconnected islands in the South Pacific (figure 5.4). One of the shortcomings of the system, Hars remarks, is that it did not have a "sincere" business plan.[60] It would always be cheaper for people in Tahiti to simply route signals through Hawai'i, where capacity is less expensive than through Australia; it would also be quicker for those in Australia and New Zealand to use the SCCN or Telstra's Endeavour, which bypasses all other islands. Therefore, even though the system would be redundant in case of a break, there was not an economic model that would make it profitable for telecommunications carriers to use it instead of existing systems, and not enough regional circulations with which they could interconnect. One cable manager tells me that the plan never "gained traction" with the carriers, as these companies were not interested in a circuitous route going through untested and potentially dangerous locales.[61] Without political interest or adequate philanthropic goodwill (in the form of economic support), cable managers doubt that they will be stopping at islands on their way, especially as long as the business model in the cable world depends on offering quicker and safer routes to be competitive. Ultimately, the lack of both insulation and interconnection kept the SPIN system from coming to fruition.

Looking at islands as nodes—with an attention to the ways that they are enmeshed in different kinds of networks that need to be interconnected—problematizes not only the network versus island schema but also the assumptions about dependency that are typically associated with networks and islands. In the epigraph to this section, the famed Antarctic explorer Ernest Shackleton describes Fanning Island as a "lonely islet" across which the Pacific Cable Board's telegraph wires extended and, paradoxically, on which they depended. There is a contradiction presented by the centrality of such a peripheral location: here, things are out of place, and for a moment the words of the North are "placed in Southern hands." This contradiction is resolved by his holistic view, in which Fanning Island reveals to him the significance of how even the smallest acts contribute to a finished plan. This chapter has placed networks in their island contexts to provoke a conceptually similar inversion: here, islands are not that which is excluded from a network; in the case of cable communications, they play a critical role in supporting the connection of mainlands, even if set up with U.S., British, German, French, or corporate support.

Nonetheless, islands must grapple with the perception of their isolation—Guam's citizens tend to perceive themselves as marginal, despite their centrality to varied transpacific flows; Fiji's industries struggle to balance an image of the island as isolated with one of it as connected; in Yap and Tahiti, images of isolation are both mobilized to secure tourists and to justify more cable systems. Yet none of the four is truly isolated. All of them remain connected to various networks: local, interisland, and regional systems. The easiest islands to network are those where cables can triangulate and leverage existing circulations. This process of hooking into existing circulations is key to achieving success as a network node, a critical point for Pacific islands that aspire to attract cables, especially those that lack the financial means to establish such systems on their own. Observing these interconnections, we can see how Guam has leveraged various circulations—of atmospheric movements, transportation systems, and military interests—to become a cable hub. We can also see how Fiji is now seeking to develop local forms of traffic that will take the place of its previous regional network. In both cases, the perception of the island as a potentially insulating structure remains important: companies continue to seek out sites where network circuits will be safe. As sites of insulation and interconnection, not isolation, islands continue to affect the topography of the overall cable network, shaping its traction for future currents.

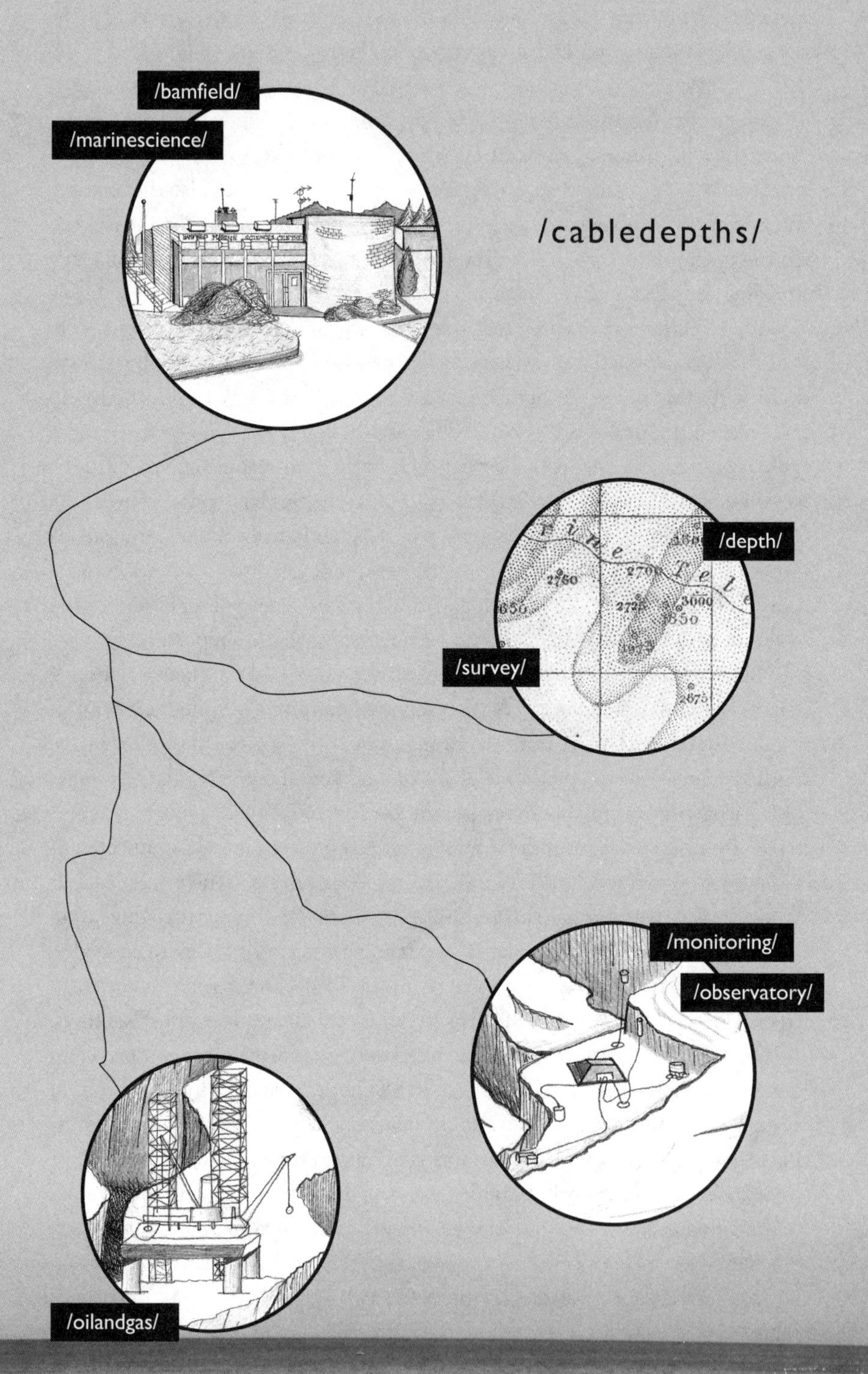
/bamfield/
/marinescience/
/cabledepths/
/depth/
/survey/
/monitoring/
/observatory/
/oilandgas/

SIX

cabled depths

The Aquatic Afterlives of Signal Traffic

Bamfield—a small village on Vancouver Island—was a critical gateway during the telegraph era, a site where international signals punctured Canada and threaded it to other territories around the world. At the same time, the station was one of the most remote places in the Pacific Cable Board's network, due to the problems in regional transportation. The nearby inlet was known as the "graveyard of the Pacific" for its numerous wrecks, and the coastal Telegraph Trail served as a lifeline for their survivors.[1] One cableman recollects that even though Bamfield was only 120 miles to Victoria, the largest city on Vancouver Island, "you might as well have been a thousand miles away for all the difficulties in transportation."[2] There was no access to the town by road until 1963, and today it remains accessible only via private infrastructures. To visit, I take a highway north from Victoria and then travel west for two hours along narrow gravel logging roads, where trucks barrel down without warning, leaving clouds of dust in their wake. When I arrive in Bamfield, I discover that there

FIGURE 6.1. Cable station turned into a marine science lab, Bamfield, Canada.

is no gas station. If I had not filled up my tank before heading out, I certainly would not have made it back.

As is the case on many islands across the Pacific, in Bamfield the small tourism industry capitalizes on the site's insularity: its remote location is a draw for visitors. As I walk around the town center, which contains only a motel, a bar, and a handful of stores, I see young people strolling along the streets, divers suiting up, and trucks fitted out for recreational camping. Bamfield would not be a destination for so many visitors if it were not for the establishment of the Bamfield Marine Sciences Centre (BMSC), which transformed the abandoned cable station into labs and classrooms (figure 6.1). The history of the Pacific Cable, which operated out of Bamfield from 1902 to 1959, is visibly infused into the space. Commemorative Pacific Cable postage stamps are displayed in the BMSC lobby, a cable memorial stands at the center of the grounds, and signs around the site describe the function of the cable buildings. The cable community originally channeled people into this remote area, justifying the upkeep of basic infrastructure and ensuring the site's continued accessibility, but the marine sciences center now serves that function for Bamfield, building on and repurposing the infrastructure it inherited.

Even though Bamfield is no longer a hub for global communications (Canada's westward cable traffic now goes south through the United States), cable networks remain critical to Vancouver, not simply for their historical impact on local infrastructure but for the region's marine scientists. Where the telegraph network once extended, there is now a large seafloor observatory, a system of undersea cables connecting scientific instruments on the bottom of the ocean. Rather than linking users on either side of the Pacific, the North East Pacific Time-Series Undersea Networked Experiments (NEPTUNE) network enables scientists to communicate with the ocean itself, tracking seafloor processes as they occur in real time and shifting the temporality and spatiality of oceanographic research. Bamfield's history reveals how marine science has built on the foundations of intercontinental communication: the undersea cable industry designed and laid the cable; marine researchers conduct maintenance on the network from the BMSC; and up the inlet at Port Alberni, the network terminates in a cable station built for a 1960s coaxial system.

This chapter moves outward from the cable station, landing point, and island to document the environmental imbrications of cables in the deep ocean, generally considered the safest part of the cable route and a layer of insulation protecting signal traffic. Indeed, the aquatic environment itself serves a medium for the signal's return. After power crosses the cable, it is routed to an ocean ground bed that grounds intercontinental currents; the ocean completes the cable circuit. However, laying cables in the deep ocean requires extensive environmental knowledge to determine where the cable's wet plant will be safest. This has entailed the coordination of ideologically dissimilar oceanic actors and institutions, from oceanographers and environmentalists to national navies and oil companies. Previous chapters have analyzed how the manipulation of physical sites and social practices shape contours in cabled environments and give traction to, though do not determine, the movements of subsequent circulations. This chapter traces how the knowledge of topography, currents, and the marine environment have been key to the extension of undersea networks and how disparate institutions and organizations, because of their shared investments in the ocean and their development of technologies of aquatic knowledge production, have shaped the deepest sections of our undersea networks.

Describing the entanglements of cables with our knowledge and inhabitation of the ocean—conducting a kind of cultural bathymetry of cable systems—I show how the creation of stable circuits of transmission is, at its core, an environmental process. I begin by documenting the cables' relationship with the mapping of aquatic surfaces. Early oceanographic research gave traction to telegraph traffic; in turn, cables shaped the archive of the ocean's depth, starting in the 1850s and continuing into the present. Although this began with simply mapping the ocean's surfaces and depths, it eventually extended to analyzing how the environment changed over time. In designing cable routes, communications companies forged ongoing institutional interconnections with navies and scientists to develop this body of oceanic knowledge. In the

1950s these actors used cables to monitor aquatic space in real time, a transition facilitated by military investment. Taking Vancouver's ocean observatory as a case study, I examine how scientific cables are facilitating new kinds of interaction with the ocean and how these developments are feeding back into telecommunications in the creation of new dual-purpose systems. The chapter ends with an exploration of the material afterlives of cable technologies and speculates about their potential alternative uses. Although undersea cables have little direct environmental impact, the history presented here reveals that they are nevertheless technologies of environmental transformation. Cables have carved out paths for aquatic knowledge production, monitoring, and extraction and have created ripples across military, scientific, and industrial practices in the ocean. In turn, they have been affected by the currents and circulations, both human and nonhuman, that unfold across the ocean. A better understanding of these linkages—of how undersea cables are literally embedded in the ocean—can help us imagine a range of new, diverse uses for cable systems and expand the uses of these networks beyond the purview of major telecommunications companies.

Mapping

Since the telegraph era, laying reliable undersea cables has required knowledge about the ocean's depth. To extend cables safely between continents, cable layers must follow existing seafloor transitions. If cables are draped across valleys, they risk being caught in currents. If they are routed diagonally down steep slopes, they might be broken by landslides. If laid across sunken ships or coral reefs, they could become twisted and tangled. Therefore, cable layers have relied on people, organizations, and technologies that produce knowledge about depth contours. It is no coincidence that the first undersea cables were laid only shortly after depth soundings began to be taken (this entailed throwing a weight attached to a line overboard and lowering it until it stopped). In the 1850s, when the first undersea cables were being laid, recommendations for standards for the collection of deep-sea data began to circulate.[3]

Oceanographic expeditions to collect depth data were leveraged to facilitate the extension of cable systems. Before the first transoceanic cables were laid, Lieutenant Matthew Fontaine Maury of the U.S. Navy was working to survey seafloor topography, generating knowledge that would help speed up marine transportation. After taking a line of readings between Ireland and Newfoundland, Maury noted that there was a plateau between the islands, "which seems to have been placed there especially for the purpose of holding the wires of a submarine telegraph, and of keeping them out of harm's way."[4] Maury viewed the seafloor and its contours in terms of their potential for cable expansion. Cyrus Field, who was leading an effort to fund the first transatlantic cable, later wrote to Maury asking about the feasibility of a cable project, knowing that oceanographic knowledge would be critical to its success. Maury shared the discovery of the telegraph plateau with Field and made suggestions about how best to lay a cable; seafloor information was integral to Field's fund-raising efforts and to the eventual success of the first transoceanic cable.

Throughout the late nineteenth century, the need for intercontinental cables stimulated a vast increase in soundings and the development of an archive of seafloor topography. Helen Rozwadowski argues that these surveying ventures, often sponsored by the American and British navies, were important in establishing powerful commercial and political motivations for oceanographic work.[5] They increased the available knowledge about the ocean, contributed to the advancement of marine science and technology, and became an important nexus of financial and military support for these activities.[6] Just as Maury did, expeditions interested primarily in marine science saw and assessed the seafloor in terms of its potentiality for cable laying. The HMS *Challenger*'s expedition in 1873–76, often regarded as the starting point for oceanography, not only made meteorological and biological observations but also conducted soundings for possible cable routes—seeing the ocean floor as a prospective cable site.[7]

The emergence of cable laying and marine science attracted popular attention to the deep ocean and helped change the way the public understood the sea. It went from being an "unfathomable barrier" to a space available to technological observation.[8] As Rozwadowski describes, a discussion about the seabed emerged in this period, including speculations about the composition of the ocean floor, illustrations of bottom sediment in newspapers, and descriptions in cable promotional material about the deep sea.[9] Prior to writing *20,000 Leagues under the Sea*, Jules Verne crossed the Atlantic on the Great Eastern steamship, interviewed crew members who had helped lay the transatlantic cable, and met Cyrus Field.[10] In Verne's novel, when the *Nautilus* visits a broken undersea cable 1,400 fathoms under the sea, Pierre Arronax takes the opportunity to describe the first attempts to lay a transatlantic cable. In popular culture, the ocean's depths were seen as a site for cable laying, and this in turn helped to legitimate continued investment in telegraphic expeditions.

As oceanography charted paths for cables to follow, cable laying and repair recirculated aquatic information back to marine scientists. Cable expeditions generated knowledge about undersea landforms, various deep trenches, and bottom conditions.[11] According to Cable & Wireless, by the mid-1930s more than 2,000 soundings were being supplied each year to the British Admiralty to correct its charts.[12] Depth soundings challenged previously dominant theories of submarine space, such as the belief that objects would sink only to a certain point before stopping, and helped establish a new epistemology of the ocean floor. An article from the mid-1930s reports that although scientists had believed that the world's ocean currents did not affect the sea below a certain depth (400–500 meters), information supplied by cable ships "led to a contrary conclusion." Ships reported great difficulty in picking up cables precisely due to deep-sea currents that altered the position of grapnels as they sank. They reported that "many strange scraps of information about the ocean bottoms" had ended up in their "chart-crammed Marine Room."[13] Technologies and infrastructures created for cable laying also came to benefit marine science. The Commercial Cable Company's *George Ward*, subsequently outfitted for ocean exploration, was used to search for wrecks and take undersea motion pictures.[14]

Cable expeditions generated information not only about the ocean's depth but also about its inhabitants. When they lifted cables from the ocean, cablemen found creatures affixed to them, providing evidence of life on the seabed. A variety of institutions documented these discoveries. The British Museum constructed a set of boxes to store marine life sent in by cablemen, and it published instructions for preserving specimens and collecting data in the *Zodiac*. In the 1920s the *Zodiac* reported that "a great number of most interesting specimens from the cables have reached our National Museum, including Corals, large stalked Barnacles, Sponges, Alcyonarians, Hydroids and Tunicates."[15] Cable crews collected specimens for the University College of the West Indies and identified whales for the National Institute of Oceanography.[16] One cableman recounted: "The rack of specimen bottles has now become a permanent feature of the *Electra*. While on cable work, various enthusiasts can be seen picking oddments off the growth on the incoming cable, and raking about in the bowls of recovered mushroom anchors and the crevices of grapnels."[17] Individual scientists, including the geologist who discovered *Globigerinella aequilateralis* in cablemen's samples, would enlist the help of cable ship's crews.[18] Others, including a scientist studying fish migration in the Indian Ocean, accompanied cablemen on their voyages.[19] Through the 1950s cable workers continued to send samples to marine scientists and helped comprise an archive of the seafloor's composition.

The extension of cables not only helped map the ocean's depths, but it also contributed to charting its surface. Before the nineteenth century, ocean navigation was a formidable challenge. Charts were unreliable, longitude and latitude were difficult to determine, and shipwrecks occurred regularly, resulting in large numbers of lives lost and relatively unreliable marine commerce. The creation of better charts, as Richard Stachurski has documented, required the exact determination of longitude, which depended on the ability to make astronomical observations in different locations at precisely the same time.[20] Prior to the extension of undersea cables, time was not standardized around the world; but after continents were linked, mapmakers could compare the times given by chronometers at two cable stations and determine the longitude between them. When the first transpacific cable was laid, it was therefore seen as "an opportunity for continuing the work across the Pacific in the interests of navigation and geography."[21] Astronomic observation huts were built on specially constructed brick or cement piers at Pacific Cable Board stations, and equipment was set up to use them for longitude experiments. By 1903, when satisfactory exchanges finally connected Sydney back to London, "longitude from the west clasped hands with longitude from the east, and the first astronomic girdle of the world was completed."[22] These interconnections—between the marine sciences, naval institutions, and cable builders—would later benefit the companies laying cables. To determine the best cable route, for example, the Pacific Cable Board submitted soundings and bottom samples to the British Museum of Natural History for examination.[23] When these actors did not work collaboratively, however, it could cause problems. Such was the case when the British Admiralty delayed the development of a transpacific line in the 1800s by withholding marine surveys and ships.[24]

The development of more expensive and higher-capacity analogue cables, the practical need to design undersea repeaters to withstand incredible deep-sea pressure, and the increasing danger of fishermen's disruptions stimulated a renewed drive toward accurate placement of cables, a heightened investment in technologies of observation, and the strengthening of these interconnections in the era after World War II. This was also linked to a broader investment in subsea technology that was motivated by naval interests in the wake of the war, as well as to marine scientific work that helped prompt a "revolution" in knowledge about the ocean's internal dynamics.[25] Military interests increasingly shaped exchanges of aquatic knowledge. Both the U.S. military and Bell Laboratories supported the making of the first comprehensive and detailed maps of the ocean seafloor, produced by Bruce Heezen and Marie Tharp in the 1950s.[26] Heezen's master's thesis drew on data about undersea cables (he studied the impact of the Grand Banks earthquake of 1929 on transatlantic telegraph cables), and he later helped Bell Labs avoid environmental hazards when routing the first transatlantic telephone cable.[27] Through their connections with communications companies, oceanographers such as Heezen were given access to private seafloor data, which shaped their final maps. Navigational technologies also increased the cable companies' dependence on the U.S. military's coverage of ocean space. Given astronomical navigation alone, a ship could not discern its exact location, especially out of view of land or in bad weather, and even through the 1960s navigational issues could prevent the precise laying of cables. However, in the 1970s this situation greatly improved with the introduction of the satellite technologies and GPS navigation, initially developed using military technologies. For the first trans-Tasman system, the planning committee suggested the temporary installation of radio navigational aids to help the cable ships, but for the trans-Tasman system in 1975, it was satellite navigation, including the U.S. Navy's TRANSIT system, that aided the ships.[28]

The focus of data production during this period turned to locating ever more precisely the cable ship and its route, newly tracking them not only in space but also in time. Precise data about the seafloor not only meant that the cable would be less likely to catch on a fishing trawl, but also that cable companies would now be able to bury the cable beneath the seafloor. Cable burial had been proven feasible in the late 1930s but was delayed by the war; telecommunications companies took it up widely with the development of telephone cables.[29] Normal deep-sea soundings could not indicate if a surface had hidden bedrock, but with the use of new technologies, cable layers could better gauge seafloor composition.[30] These included echo sounders that bounced sounds off of the seafloor to determine its depth (improving on earlier line surveys that were incapable of providing continuous measurement) and the sub-bottom profiler, which could penetrate mud and provide an image of subsurface strata.[31] With the side-scan sonar technique, the survey ship towed a vehicle behind it to generate a wide profile of the seabed and assess "suitable cable routes" through undersea mountain ranges.[32] Cable companies also determined seafloor temperature and used current meters to determine the strength, direction, and velocity of subsea currents.[33] By 1967 the undersea camera was an essential tool. One engineer commented: "To the trained eye, photographs of the sea bed yield a great deal of information on the nature of sediments, bottom currents and other facts likely to affect the security of a proposed cable. . . . The knowledge gained adds to the store of oceanographical data accumulating for everyone's benefit."[34] The companies' seafloor technologies were used to assess not just depth but also a variety of seafloor processes—from composition to currents—that were best described over time.

Cables gained a new function in many of these seabed-assessment technologies: they linked the ship directly to instruments and remotely operated vehicles (ROVs).[35] By the 1960s cable companies' ships could tug a "survey sled" that was equipped with lighting and a TV camera and was remotely controlled by people on the ship.[36] An undersea "control cable" enabled cable crews and oceanographers to study the nature of the seabed in real time.[37] The U.S. Navy was a key investor in ROV development, as it began experimenting with using cables for its Cable-Controlled Underwater Recovery Vehicle in the 1960s. These ROVs helped pick up the detritus of war, including torpedoes and bombs, from the seabed and aided in recovery missions for manned submersibles.[38] By the 1980s ROVs had become one of the most important tools for assessing the ocean floor and were making cable maintenance more feasible and inexpensive, further uniting the cable industry and oceanography.[39] SCARAB (Submerged Craft Assisting Repair and Burial), released in 1981, could use a powerful water jet to blow a hole in the seafloor, extract a cable, cut it with a circular saw, and attach it to new lines so that it could be lifted to the ship. Once all this was done, it could lay the cable again—all at depths of up to a mile. One Cable & Wireless submarine systems engineer explains: "It's the first time . . . that we've been able to put cable technicians, as it were, on the sea bed. They work directly through remote devices, the TV system becoming their underwater eyes."[40] Prior to the use of ROVs, cable companies could retrieve the cable only by dragging a grapnel along the ocean's floor. Although SCARAB was initially constructed on behalf of the cable industry, it used existing oceanographic tools and could also undertake other kinds of underwater activities, such as locating and salvaging aircraft wreckage.

Oceanographic institutions continued to be a part of the exchange of undersea data and technologies. As in the telegraph era, cable laying still relied on information and consultations from marine scientists and oceanographers, and the production of a subaquatic archive. During this period, AT&T contracted with the Lamont Geological Observatory to gather information about the seafloor, and oceanographers repeatedly went to AT&T to consult the engineers.[41] The managers of the Commonwealth Pacific Cable (COMPAC) cable relied on advice from geophysicists on potential seismic activity near the route. At times, cable workers were invited to travel on marine research ships.[42] In turn, the cable companies' developments fed back into oceanographic work. Cable survey expeditions continued to find and then chart new landmarks. Biologists went to Bell Labs to study old telegraph cables, in attempts to determine the cables' effects on ocean life. In the late 1960s the U.S. Underseas Cable Corporation offered a new line of undersea television equipment that could be used not only for cable laying, but also for marine biology observations.[43]

Today the work of cable laying is performed by a small number of cable suppliers and specialized cable survey companies, such as Earth Science and Surveying (EGS), who continue to depend heavily on these archives of seafloor topography and a reservoir of available knowledge. One employee reflects on Cable & Wireless: "Because of its long experience it has built up comprehensive worldwide records of laying and fault data for both coaxial and telegraph cable systems. This provides an invaluable and unique information base from which to work when new cable routes are being planned, particularly when a new route closely relates to an existing system. . . . ANZCAN will, in a general way, follow the route of the COMPAC system, for which Cable and Wireless has full survey data."[44] The survey is critical for the line's stability; engineers believe that the better the map, the more durable the system. Existing environmental data are always the starting point, and even though companies always redo surveys using the latest technologies, this often is used to prove the feasibility of an existing route. The interest in precision has evolved to the point where computers on today's cable ships continually index the ocean beneath them, assessing the currents, turbidity, and a number of other features; determine the speed at which the cable is paid out; and automatically steer the ship so as to control exactly where the cable lands on the seafloor. Dave Willoughby, a cable route engineer, tells me that many organizations want to share information about the seabed, but it is valuable, and his company typically turns down requests to do so: "We weren't going to make it publicly available. It would be giving competitive advantage to other people: where our systems are, where our systems are planned, but also the logic that goes into planning a route."[45] At times, companies do turn information over to others within established circuits of exchange, including national geological and hydrographic departments and other cable companies, on the basis that it will remain confidential.

Much of this knowledge continues to be stored inside the walls of these companies, passed down informally from generation to generation via on-the-job training. Willoughby explains to me the importance of existing seafloor knowledge to determining a cable's route. When he arrived at AT&T Submarine Systems in the early 1990s as part of an influx of a "new breed" of younger workers, long-standing employees had been doing things in the same way for the past twenty years. The process would begin with a notice from their sales team that a customer was interested in laying a cable between two points. The engineers would go to a paper catalog, make a list of the charts they needed, and then order them, a process that took five to ten days. Companies conduct an extraordinary amount of marine engineering before even going into the water; this includes poring over old cable surveys and charts from national oceanographic departments. On these charts, the engineers would mark potential routes with pencils. Many of their concerns date back to the telegraph period: they want to make sure that they do not cross unstable zones, run the system parallel to sloping areas, or route it over deep canyons, which would suspend the cables in the water. This requires extensive knowledge of seafloor topography and composition, and draws on Willoughby's background in oceanography: "It was all human knowledge. . . . My co-workers who had been doing work for five years would say, 'Hong Kong–Guam, we've done that already.'"[46] Willoughby could take an existing documented route, or he could propose a new one. "There was a lot of personal judgment that went into it," he says, "and that's also how you learn." He recounts the kinds of discussions he and his colleagues would have: "I might have said, 'Well, it looks like you are taking the long route around. Here's a straighter route.' And that's where the guy who's more experienced would say to me, 'Well you see, you're crossing on a steep slope area. If there's a mudslide, your cable's going to have problems, so this is one case where it makes sense to take a longer route.' And there might be another case where I'd be advocating a longer route and the experienced guy said, 'Yeah, but you just added 100 kilometers onto the route, that's millions of dollars.'" Dave remarks that since "no one goes to school to become a cable route engineer," much of his job has involved "learning by doing."[47]

The process of developing undersea routes has changed remarkably with the advent of digital technologies. Engineers have moved from paper charts and hand calculators to personal computers with graphical geographic information system interfaces that enable them to manipulate potential routes in real time. They no longer wait days for paper charts but look at a screen that has continents, bathymetry, bottom features, wrecks, and active and obsolete cable systems. Computation accelerates the process of cable design, as engineers can mock up, circulate, and discuss potential routes much more quickly, and sales teams can give more detailed and immediate feedback. This process opens up cable laying to new actors and allows aspiring cable builders to prepare their own diagrams. One cable entrepreneur describes his use of Google Earth and the British Hydrological Society website to explore potential cable routes: "You can put your mouse over a point on the ocean, and down the bottom it'll tell you the depth, so I've used that to search where to put branching units. Look at the depth of a possible route into the island . . . you click on a place, and it sits there going z-z-z-z-z and suddenly whatever detail there is of the seafloor is there."[48] The broad dissemination of aquatic information and the ability to access the ocean's depths have not lessened the importance of the marine engineers or the stocks of cable route information in the world's largest cable companies. According to Willoughby, "It still is a specialty—there is still some art, craft, and science that goes behind it."[49] He says that although it is easy for people to go to Google Earth and say that they know the cable route, it is only that easy "until something goes wrong. The people doing cable route engineering are still the same type of people that were doing it twenty years ago. . . . They still need to know the science of the seafloor."[50]

The technological spanning of the ocean via undersea cables enabled the mapping of the seafloor, its inhabitants, and the ocean's surface. From the telegraph period on, a set of institutional interconnections among marine scientists, cable companies, and national navies facilitated the sharing of knowledge about the ocean's depths and made it more feasible to lay cables in the region. During the Cold War, naval interests increasingly mediated the boom in oceanographic and cable development. New deployments such as echo sounders, underwater television cameras, and ROVs gave scientists and cable layers access to seafloor processes over time, propelling a critical shift in the use of undersea cables for monitoring and scanning the ocean's depths. Broader military and scientific investment in ocean technologies coincided with the cable companies' need to lay and bury cables with precision, tying these practices tightly together.[51] The laying of cables today continues to depend on an extensive knowledge of the environment that is stored by a small number of companies and that has been built on a history that cable companies, navies, and oceanographic institutions share. Just as the interconnections at the cable station and the strategies of insulation at the cable landing, these circuits of seafloor knowledge sustain and solidify particular transoceanic contours, giving undersea networks traction in aquatic environments.

Monitoring

Coinciding with the shift toward more precise monitoring of the seafloor in the era after World War II, the U.S. Navy developed another use for cables that mobilized them as technologies of underwater detection. Cables had long been able to register undersea processes. Early telegraph cables could pick up signals generated by the earth's magnetic currents, lightning, and other phenomena as they passed under the ocean. However, these noises were coded as interference. During World War II, given the increasing importance of submarines to warfare, the U.S., British, and Soviet Navies began to realize the significance of identifying and recognizing signals emanating from the depths; this led them to install acoustic arrays in shallow waters near harbors to detect the presence of an enemy.[52] Beginning in the early 1950s, the U.S. military developed a secret network of undersea cables, the Sound Surveillance System (SOSUS), that served as listening posts for Soviet submarines. The military contracted AT&T to develop the system, further drawing from the company's extensive knowledge of the seafloor and cementing its relationship to defense.[53] Cables extended out from monitoring stations on U.S. territory and used hydrophones (underwater microphones) and passive sonar to listen to the ocean's soundscape. These cable-supported tools could locate submarines 2,000 miles away.[54]

The first SOSUS arrays were installed on the East Coast of the United States in the early 1950s, and Pacific arrays became operational in 1958. These systems recorded all kinds of noise, and U.S. Navy operators were trained to spot the signatures of enemy submarines and ships in the incoming data. At times, the arrays were used for nonmilitary purposes, such as assisting U.S. Customs officers to track vessels suspected of drug smuggling.[55] The geography of these systems in some places mirrored AT&T's commercial systems: shore terminals, termed Naval Facilities, were often established in places where transpacific cables had landed, such as at Hawai'i and Midway Island (near the landing point for analogue cables) and Guam (near the original telegraph landing point). Another network, the Missile Impact Location System, was also installed in the late 1950s to determine where test missiles landed.

As telegraph cables were retired from commercial service in the 1950s and 1960s, another group began to use cables to monitor the ocean. In this case, it was scientists who hoped to use retired systems to gather marine data. When the COMPAC cable went live, the telegraph cable it replaced was made available to teams of scientists to make geophysical observations of the seafloor.[56] A group from the University of Newcastle studied the movement of water in the ocean's depths by measuring the voltage differences between points along the cable. Stanley Keith Runcorn, from the University of Edinburgh, used these cables for his research on continental drift and tectonic activity. In 1957 a physicist from London University and a team from the University of California set up a station at Fanning Island to study geophysical processes in the Pacific, taking earth current measurements from Fanning's cable termination.[57] When the Fanning Island station closed in the 1960s, it became the location of the Hawaiian Oceanographic Institute's Pacific Equatorial Research Laboratory (PERL), a nonprofit organization that conducted meteorological observations, research on equatorial currents, and other oceanographic work (and has since contributed to the global database on climate change research).[58] Martin Vitovsek, who set up PERL, developed instruments to place at the end of cables that gave deep-sea measurements and could further the scientific understanding of tsunamis.[59] Two abandoned cables from the Commercial Pacific Cable Company were also handed over to the Hawai'i Institute of Geophysics for this purpose.[60]

The transition from analogue to fiber-optic cables in the late 1980s brought similar opportunities, and the investments that telecommunications companies had made in particular aquatic environments were transferred to scientists. Although many old cables were still functional, maintaining them was no longer economically viable and they became commercially obsolete. Scientists, especially in the seismic research community, sought to harness undersea networks' potential for monitoring the oceans. Responding to requests in the United States and Japan, in 1991 AT&T and the Kokusai Denshin Denwa Company (KDD) donated a section of Trans-Pacific Cable 1 to scientists. In 1996 the Hawai'i-2 undersea telephone cable was also donated, and two years later the Hawai'i-2 Observatory (H2O) was installed at its end.[61] Initial instruments in this cable observatory included a seismometer and a geophone (two devices for measuring ground movements), a pressure sensor, and a hydrophone.[62] The donation of these cables benefited not only the scientific research community but also the cable companies, which were spared the cost of retrieving and disposing of unwanted cables. These cables still terminate in the original Cold War–era cable stations, a process made possible by the relative openness of the stations today.

Toward the end of the 1980s, the U.S. Navy found it more difficult to bear the expense of operating and maintaining the SOSUS arrays. Scientists in the National Oceanic and Atmospheric Administration (NOAA), who had known about the network for at least a decade, became interested in it. Verena Tunnicliffe, a marine scientist, recalls: "The guys in NOAA, the geophysicists, knew about the fact that the Navy was getting signals. Every now and then they would have heard about some neat event that the Navy had heard. Then they started the discussion: 'Okay, you guys have got a cable and you're hearing things offshore. Is there some way we can connect in so we can hear just the things that are interesting for science?'"[63] The idea of opening the SOSUS network had been floated for some time but never pursued. In 1978 Harvey Silverstein had suggested that a global ocean monitoring organization be formed, which could be valuable to scientists as it could "monitor and follow movements of everything from shrimp to whales and, consequently, observe some of the secrets of marine creature migration."[64] In the early 1990s, at the request of NOAA scientists, the Navy finally began to send signals from the SOSUS cables to the acoustic monitoring project of the Pacific Marine Environmental Lab. The network was routed in a secure room, and sensitive information was filtered out.

Marine monitoring via undersea cables represented a paradigm shift for deep ocean science. Previously, data from the seafloor had to be gathered via trips out to the middle of the ocean, a difficult and expensive endeavor. Tunnicliffe described the process to me: "We could only get out in the summer in the northeast Pacific. The winter's too bad, so every summer we would be going out there, and there would be some big change—like clearly some lava had poured out, or there had been a big spreading event. At that time people were saying, 'If only we could connect to this place during the winter or fall.' So that desire to have full-time monitoring, and knowing when an event had happened, had been there in the community all the way back into the '70s."[65] As a result, scientists have observed that even with satellites and oceangoing vessels, "The record of land-based measurements contrasts starkly with the near-total absence of long-term geophysical data from the seafloor."[66] With the hydrophones of the SOSUS network, scientists could continuously monitor animals (for example, whale communications and migrations) and geophysical processes (such as underwater earthquakes). They could also use SOSUS to respond to events. For example, immediately after the Pacific Marine Environmental Lab began to get real-time data, scientists heard an eruption on the Juan de Fuca ridge in the northeast Pacific. Unlike in the days of periodic ship-based journeys, the team was able to quickly mobilize a crew, visit the site, and gather data that documented the immediate effects of the volcano.

There were still limitations. Scientists had to stay close to existing SOSUS routes, so scientific knowledge followed the contours of military geographies. Tunnicliffe argues that they still could not get near enough to undersea events: by the time they had assembled a crew, three weeks would have passed.[67] This delay made using the network particularly difficult for microbiologists or chemists for whom studying the events' immediate effects were critical. Scientists began to imagine the possibility of laying their own instruments, ones that were not subject to the Navy's filter, and developed plans for their own seafloor observatories. In 2006 the Victoria Experimental Network Under the Sea (VENUS) was launched to monitor processes on the coastal seafloor of British Columbia. In 2007 the Monterey Accelerated Research System (MARS) was laid off California's coast. NEPTUNE, a large-scale ring off the coast of British Columbia, went live in 2009 and was designed to cover different kinds of tectonic areas; major ocean ecosystems; and types of nutrients, habitats, energy sources, and life-forms (figure 6.2).[68] Following NEPTUNE, cabled observatories were proposed and set up around the world, including in Hawai'i and Taiwan.[69] Rather than relying on existing cables, these projects hired telecommunications companies to install state-of-the-art cable systems on the seafloor.

FIGURE 6.2. Artist's rendition of the NEPTUNE undersea observatory. Image by Juliane Richter and Birte Wagner, GEO, courtesy of Ocean Networks Canada.

These cable observatories did not displace, but were compatible with, existing forms of ocean observation that used ships, buoys, and satellites. Due to the uneven temporality of biological occurrences, Mairi Best, Christopher Barnes, Brian Bornhold, and Kim Juniper argue, ship-based observations, with their limited sampling conditions and sporadic data capture, might confuse longer-scale events with one-time occurrences.[70] Ocean observatories shifted the temporality of the ocean, mobilizing the techniques developed by SOSUS and the ROV that used cables to regularly capture data from seafloor sites. With specially designed junction boxes, scientists can put down a range of instruments (from video to current monitors), index different kinds of processes, and input many types of data (geophysical, biological, chemical, and so on) at once. NEPTUNE's builders argue that this gives it the potential to facilitate interdisciplinary collaborations between scientists who are interested in different ocean processes. By indexing multidimensional episodic changes as they play out, the builders hope to be able to study transitions in a new way, one that can better account for the ocean's complexity.

Not surprisingly, NEPTUNE's creators draw on the imaginary of a liberatory cyberspace to portray it as a site of disciplinary interconnection. Martin Taylor, the president and CEO of Ocean Networks Canada, told me that "wiring the oceans, enabling power—electrical power—and the Internet to liberate our ability to study the oceans was the concept from the beginning."[71] Taylor believes that extending the Internet to the deep sea is liberating not only since it offers new opportunities for interdisciplinary work, but also because marine science research can now extend globally: a scientist does not have to be on a ship or near an ocean to conduct experiments. Yet, as Stefan Helmreich observes, this "networked ocean" ultimately "circuit[s] through the resistances and capacitances of landed politics."[72] NEPTUNE remains limited—as are all cable networks—by the resources of the community that maintains it. Scientists rely on specialized oceanographic research vessels for upkeep. The vessel that NEPTUNE uses is booked a year in advance, and all maintenance is conducted on two yearly trips, on which crews look for signs of corrosion; clean, inspect, and recalibrate instruments; and collect sediment samples. Although instruments could be built to last for the duration of the cable's life if they were made out of the most expensive materials, finding a supplier to build instruments within budget is difficult, and as a result extensively corroded ones have to be thrown away. If an instrument breaks immediately after a maintenance trip, the scientists who use it must wait until the next trip to reinstall it. Just as the work of cable operators has been critical to international networks, these maintenance trips are key to the operation of NEPTUNE. Without routine servicing, the network's functionality would degrade.

Although NEPTUNE "frees" marine scientists from some historical constraints, it also imposes new kinds of spatial and temporal forms on them and the processes they study. The nodes and instruments that reside on the seafloor inevitably affect this space: they generate heat that can alter the aquatic environment. As scientists who use the system gain online access to it, they are also displaced from ocean trips themselves. They no longer simply drop their own instruments but install them on a network with numerous cable connectors, interfaces, and maintenance schedules that are beyond their control; they must collaborate with others in logistical ways as well as intellectual ones. The cable observatory also presents new kinds of technological challenges. NEPTUNE's scientists had to develop a set of undersea power converters to transfer the voltage from the cable to the instruments, an information and communication technology architecture to process the signals, and a system that enabled scientists to work remotely with cables in the ocean. These challenges have turned into commercial opportunities. New node and power conversion technologies can be sold to other ocean observatories and to companies in the oil and gas sector. As has been the case for cable networks since the 1960s, NEPTUNE is completely dependent on the cabled ROVs for much installation, repair, and maintenance, all of which come with significant costs. The maintenance of these systems thus benefits from and feeds into the larger set of investments in the exploration of the ocean that was pioneered by navies during the period after World War II and by the oil industry in the 1980s.

The historic capabilities of the telecommunications industry were key to establishing NEPTUNE. Its builders contracted with Alcatel-Lucent Submarine Networks, a major cable supplier, to lay the network. The industry's high level of design engineering means that the scientific research community has a reliable network that will last for decades. Like many commercial systems, it has a recognizable loop configuration (off which extensions branch to nodes); if the backbone goes out, Alcatel-Lucent's cable ship must be called in to make repairs. Observatory cables face the same problems as commercial cables, especially threats from fishermen. However, the unique needs of ocean observatories challenge the typical operating procedures of the cable industry. If a fisherman strikes a node, the cable ship would have to engage in the retrieval of this large object, a process that pushes its existing capabilities.

Building cable observatories entails a culture shift for the cable industry, which has focused entirely on safety, security, and keeping the cable insulated from the ocean. This operational approach comes into conflict with scientists' attempts to connect to the ocean. For example, during the planning phase, the NEPTUNE network intentionally was laid close to the ridge zone, precisely the area that cable networks have avoided. Tunnicliffe describes the interactions between the cablemen and scientists as based on "two completely different cultures." For example, on the cable ship, scientific instruments sat out in the open, as they would on the seafloor. Tunnicliffe remembers: "The instruments were all sitting exposed because we had to measure the seawater conditions. And these guys on the ship were just like, 'You guys are nuts, don't you know anything about seafloor conditions? Let's slap some metal over the whole thing.' I said, 'No, we want to measure ocean water and you can't slap metal over the top.'"[73] Scientific cables challenge the typical approach to insulating subsea technologies precisely because they are meant to interact with marine life. Tunnicliffe continues: "So just where we wanted to lay the cables, they go: 'No, you can't do that for exactly these reasons.' And we say: 'The reason you don't want it to go there is exactly the reason we *do* want it to go there—because that environment is hugely active, it's got huge currents, the big sand dunes.'"[74] The network's geography is a compromise: the backbone is kept away from the dangerous areas, and the extension branches with instruments are allowed closer—NEPTUNE thus balances insulation from the undersea environment and interconnection with it.

The use of undersea cables for science builds on the legacy of military investment in cable monitoring, over a century of technological development in the cable industry, and contemporary ideas about digital media. In giving scientists and publics a way to access and manipulate the seafloor, ocean observatories constitute a new source of demand for fiber-optic undersea systems and introduce a new potential for cables beyond simple observation: they might be used to interact with the deep sea. Tunnicliffe describes the transition this way: first, "this was the ocean talking to us. Then you get to the point of we want to talk to the ocean, the other way. How do you put things out there that you now control?"[75] She suggests that the next step might be an active engagement with the ocean, including setting up experiments on the seafloor. Best speculates about the possibilities of interactivity: "As we hear that there is an earthquake going on, we can turn on the camera lights in midocean. We can go there with the mobile crawler and see something. That's really very powerful. [We can] respond and react to things that are going on in real time, anywhere in the world."[76] Although observatories remain a small market, just pocket change compared to international telecommunications and the oil and gas industries, the information they provide about deep-ocean dynamics, the institutional interconnections they pioneer, and the new uses of undersea networks they develop continue to shape the conditions for cable laying.

Expanding Uses

As a result of these successful deployments, conversations about science cables have migrated back into the commercial telecommunications industry. Recently there has been a move to consider "dual-purpose" undersea cables that might benefit both scientific and telecommunications communities, which would be only the latest in a long-standing series of collaborations between them. These cables would be constructed by putting sensors on the repeaters of new transoceanic cables as they are laid. The sensors could collect a range of data about, for example, the movement of ocean currents; levels of oxygen and greenhouse gases; seismic movements; temperature; geophysical, biological, and chemical data; and even underwater video. With only a slight modification of the existing system, proponents of dual-purpose cables argue that companies could build a network that would monitor for tsunamis and rises in sea level.[77] Just as the advocates of ocean observatories do, promoters of dual-purpose cables describe the oceans as a place with "poor connectivity" that limits the quality of observations that can be made.[78] Dual-purpose cables, so their promoters argue, would give scientists access to broadband at remote ocean locations.[79]

In a talk at the Pacific Telecommunications Council Conference in 2012, Ekaterina Golovchenko, managing director for TE SubCom, a leading cable supplier, gave what she called "a controversial presentation" on dual-purpose cables.[80] Expanding the uses of undersea cables appears at first to be a win-win situation in the industry: sensors could turn all undersea cable networks into scientific observatories with little modification, increasing cable owners' knowledge of the aquatic environment, diversifying their sources of revenue, and potentially contributing to climate change research. The presentation seemed fairly straightforward and not at all controversial, until the question-and-answer period. Then one of the audience members stood up and asked: "Could these be used for military or defense purposes?" Golovchenko hesitated, then responded that she was just there to deliver the information about the technologies, and it would be up to the audience to determine what purposes the data could be used for. Military cables had once been mobilized for science, so could these science sensors be used for military purposes? Given the historical connections between these industries, it was certainly possible.

Making commercial transoceanic cables into remote sensing technologies raises concerns about sovereignty, as they could hypothetically be used for resource exploitation or military surveillance. As a result, the undersea cable industry is reluctant to expand the functionality of its systems. As the regulatory expert Kent Bressie observes, international law grants undersea cables unique freedoms that other marine activities do not enjoy. Marine data-gathering activities in particular are subject to many more restrictions, and jurisdiction over this data collection has been "hotly contested" for years.[81] Cable survey ships typically receive greater freedom than general hydrographic surveyors, who are more extensively regulated and controlled by coastal states. Given their multiple purposes, it is not clear how dual-purpose cables would be regulated. The industry fears that if companies put data-gathering sensors on the systems, they might have to relinquish certain freedoms.

The interconnections between actors invested in the ocean—from marine scientists to militaries—have more recently expanded to encompass another set of potential users: the extractive industries that explore and exploit the deep sea's resources. Economies of extraction have long been important to the development of communications networks. At first this was because cables were often laid to sites that were being economically developed. For example, the cable in Port Alberni, Vancouver, came ashore by the Alberni Plywood Division (today NEPTUNE extends under a barrage of logs floating in the Alberni Inlet). Smooth transportation by road to the Bamfield Marine Science Center is possible only with the help of the logging industry, which in turn requires communications. Offshore oil developers have been a market for undersea cables since the coaxial era. One ad from a 1969 issue of *Underseas Cable World* boasts: "Practically all communications requirements associated with offshore drilling rigs and production platforms can be met with modern submarine cable systems engineered by LANDCOM."[82] Indeed, it was the offshore oil industry that first made extensive use of commercial cabled ROVs, often taking advantage of prior developments in military work.[83] As they extended further and further into the sea in the 1980s, offshore platforms become more dependent on effective communications to manage their activities. In 1997 engineers from AT&T's Submarine Systems described a trend in offshore oil production toward fewer and larger offshore "host" complexes, which were then linked to smaller platforms. The engineers reported that "host facilities tend to be highly complex production platforms that warrant a high quality, high capacity communication system," including undersea cable systems.[84] In 1999 Petrobras, a leading company in offshore oil production, extended a fiber-optic network to six oil platforms off the coast of Brazil. Other systems were deployed in the North Sea, off the coast of Azerbaijan, and near Qatar. In the Gulf of Mexico—spurred in part by communications failures during hurricanes Dennis, Katrina, and Rita—British Petroleum extended the Gulf of Mexico Fiber Optic Project to several oil platforms.[85] Like the NEPTUNE network, these projects rely on the undersea cable industry for construction, maintenance, and technology development. Cable technologies have not only helped produce marine science that gauges global warming, but also, via their use in the oil and gas sector, continued to feed back into climate change.

When they provide regional communications, these systems also function as dual-purpose cables, boosting both network access for terrestrial users and oil production. A paper presented at the 1997 SubOptic conference argues that the developments of offshore complexes "create a natural opportunity for cooperation between the offshore and onshore telecommunication industries to install and co-utilize an undersea cable network."[86] Some have suggested that offshore platforms could be used as termination places for international cables—a kind of cable station.[87] In addition to the oil industry, other ocean users, such as wind-energy developers, have linked their coastal installations using undersea cables. These users now have to tackle the same issues that cable companies have dealt with for a long time: seabed mapping, security, and the negotiation of coastal and aquatic spaces with other inhabitants. They often have to deal with the telecommunications-naval-oceanographic complex to navigate a densely used coastal zone.

The U.S. Navy also continues to use undersea cables; but unlike the oil industry or scientific observatories, it keeps its systems (informally termed *black fiber*) secret and rarely opens them to the public. The continued expansion of different kinds of cable systems, both single- and dual-use, has led to the need for more coordination on the seabed and an increase in regulation. In the cases where different companies are planning cable systems to the same endpoints, the situation remains first-come, first-served—a process that still forces ideologically dissimilar actors to undergo informal dialogue and requires companies to let each other know about their routes in advance. For the Navy, which does not post its charts, this is coordinated via the Naval Seafloor Cable Protection Office, which individually contacts cable companies and asks them to route around specific zones. Prior to the 1990s and the expansion of these groups of users, this could be handled easily, especially given the tight links between SOSUS and AT&T. Employees often passed fluidly between companies and the military, and these social connections helped prevent conflicts between the various cable-laying endeavors. Following privatization, deregulation, and the expansion of cable uses, a representative from the Naval Seafloor Cable Protection Office observed that this "social network thinned out."[88] The monopoly carriers no longer dominate the market, and the companies that are expanding the uses of cable systems often do not have experience coordinating with other networks.

As they move through the deep sea, intercontinental cables become intertwined with a wide range of actors invested in undersea networks. Long-established connections between the Navy, oceanographers, and the cable industry have produced knowledge about and technology to negotiate the contours of the seafloor, making it feasible to lay cables there. Now the emergence of digital technologies and dual-purpose cables has enabled new potential users to take part in process. The diversity of companies involved in cable laying is growing, including a range of extractive industries (such as deep-sea mining), renewable energy technologies (from wind turbines to tidal power), and ocean observatories. These companies provide a set of alternative imaginations about what undersea cables might do beyond carrying traffic between nations. Marine scientists imagine cables as a site for experimentation; climatologists see them as a site where global Internet growth can be leveraged to document climate change; oil companies see them as a technology to enhance the reliability of resource extraction. During the consortium era, sharing information would have been relatively easy, as many national telecommunications companies and cable ships were owned by their government. Today, however, the dissemination of such information is increasingly contested, as data become subject to additional regulations and spark questions about issues of sovereignty. As a result, the breaking apart of historical monopolies and the expansion of cable uses is wearing down the groups' tight interconnections, both in these exchanges of data and in the coordination of route planning.

Recycling

Possible alternative uses for the network can be found in artistic imaginations. The artist Renny Nisbet's audio sculpture *Soundings* consists of a single speaker that broadcasts continually changing sounds into the Porthcurno Telegraph Museum's garden. These sounds have been gleaned and remixed from signals still emanating from an old telegraph cable that once connected the area with Spain. A sign in the garden tells us that we are hearing "the cumulative sound of the Earth's magnetic field, atmospheric radio emissions from lightning, as well as man-made electromagnetic noise." *Soundings* directs our attention toward the persistence of cable systems long after they have fallen into disuse, and it configures the cable as a potential source of artistic content. John Griesemer's novel *Signal & Noise* points us to another alternative reading of cable data. In the book, a character named Otis Ludlow spends much of his time at a cable station in Ireland, monitoring a broken transatlantic cable. Although the other employees perceive the shifting levels of electricity as noise, Otis documents messages that come through the broken cable, a cord that he is convinced connects them to a "greater world" and allows the spirit world to communicate with him. He writes in his journal: "The entire ocean speaks to us through the cable. Its storms and magnetisms beat out a message. We have given meaning to certain patterns, but who is to say there aren't patterns in the rest of what flickers through the light."[89] Cables have always been a potentially multipurpose interface with the ocean, a place where we could access different kinds of signals and come to know the ocean itself. The more diverse uses we have for these systems, the more kinds of people can engage with them, whether for intercontinental communication, climate science, or the arts. The diversification of these systems is one of the best ways to expand and democratize the use of undersea networks.

When cables cease to be commercially viable, they remain in the ocean and create various ripple effects through organizations invested in marine space. After cables are disconnected and decommissioned, they often disappear from view. Early cables were left on the seafloor after they broke or became obsolete, their locations often removed from cable maps and nautical charts. Today cables can also be deinstalled and retrieved or rerouted; this is done more often to keep the cables from conflicting with potential alternative uses of the seafloor than to avoid any documented environmental impact.[90] In fact, pulling the cable up from the ocean can disrupt the seabed and aquatic creatures that have fastened themselves to the cable, especially if it has been there for decades. Although companies are not always required to pull up undersea cables, they are not allowed to abandon ownership or responsibility. One engineer tells me that when his colleagues were trying to bury a cable off the Sydney coast, they had to remove an old telegraph cable that was in the way. Before they could do so, they had to track down the cable's owners—a difficult task, especially given that these systems could have been laid over a century ago. He tells me that one of the biggest problems in shore landings is determining what other cables exist and making sure that cable layers do not cut someone's active cable.[91]

The reuse of the SOSUS system is just one example of recycling of a cable network. In some places, obsolete cables are being relaid to connect uncabled islands. For example, the PacRim West cable was retrieved from the seafloor and used for a new Australia—Papua New Guinea route (APNG-2). Over the past twenty years, another possibility has been to use old undersea cables to build artificial reefs, where cables generate new habitats for marine life. The coastal zone off Ocean City, Maryland, is home to one of the largest cable reefs in the world (figures 6.3–6.5). During the cable boom in the late 1990s, AT&T began to decommission old cables; environmental groups' protests and directives of the Federal Communications Commission forced the company to retrieve many old cables from the seafloor. The company partnered with the Maryland Artificial Reef Initiative and the Ocean City Reef Foundation to deposit coils of used cable 50 to 300 miles long into the ocean to create Cable Wire Reef, which is now a popular spot for fishing and diving. A representative from the Ocean City Reef Foundation tells me that these cables are one of the best habitats for mussels the organization has ever had.[92]

The cable recycling process, like the conflicts at cable landing points, is geographically and culturally specific. Although environmental groups in California created interference for the commercial extension of undersea cables, in Maryland such groups facilitate the cleaning and redistribution of these cables to the ocean. A similar cable recycling plan was proposed in New Jersey, and the Isle of Man has conducted a survey and assessment for the potential development of a cable reef.[93] The process has not gone so smoothly in other places. As I discussed in chapter 2, when the Vietnamese government allowed

FIGURES 6.3, 6.4, AND 6.5. Cable Wire Reef, Maryland. Courtesy of Ocean City Reef Foundation.

fishermen to pull up unused cable, the fiber-optic systems were disrupted and the recyclers were prosecuted.[94] In the United States, cable companies with precise knowledge of the systems and good coordination with government and scientific organizations carry out the retrieval of undersea cables. In Vietnam, however, the failure to provide adequate information about the cables and their location—demonstrating the lack of connection between those who lay and those who retrieve systems—means that old cables will probably remain on the seafloor for some time.

As this chapter has demonstrated, undersea cables are versatile technologies, extended not only between landmasses but also from the shore to offshore oil platforms and from on-land science centers to instruments in the ocean. Looking at undersea cables as environmental technologies not only entails looking at their sedimentation along the routes of previous systems, but also at the ways that they have affected the history of ocean science, oceanography, and the militarization of the oceans. Oceanographers detected contours on the seafloor that cables could follow, and in developing their systems, telecommunications companies conducted informal explorations of these depths that provided data for marine science. These varied efforts and initiatives have carved out channels in the seascape, making certain pathways more affordable and certain routes more achievable. Undersea cables not only reflect the history of commercial telecommunications but have played a significant part in expanding our underwater activities, our knowledge of the ocean, and our abilities to monitor its processes. Since the nineteenth century, what we know about nature, ecology, and the ocean has been shaped by cable technologies. Transoceanic communications networks that support today's Internet transactions continue to extend in relation to the existing contours—cultural, technological, and epistemological—of the ocean. Experiments with nontraditional uses of cables prompt us to consider the range of their alternative potential purposes—whether environmental, commercial, or artistic—and to think critically about the directions in which to channel their currents.

CONCLUSION

surfacing

Every once in a while, in the midst of stormy seas, heavy tides, and eroding sand, an old telegraph or telephone cable surfaces on a beach. Wandering along New Zealand's west coast, I find a coaxial cable that emerges from a sand bank, extends over a small pond of water, and submerges back into the beach before it heads to Sydney (figure C.1). I take off my boots, wade into the pond to touch it, and wonder for a moment at its bare materiality: its eroding armor, still strong copper core, and the aquatic creatures that have made it home. I have seen cable surfacings in local news and circulated on Internet forums, which often use the cable to playfully reflect on the persistence of history (figure C.2). In 1977, 1997, and 2008 there were reports that the Doubtless Bay cable in New Zealand had surfaced, and pieces of it have been cut out for the regional museum. These stories rarely extend beyond an initial "Aha!" moment. Like the cable, they appear out of nowhere, and they soon slip back into the oceans of our contemporary information sphere, failing to link to the cultures, practices, and environments that cables shape and inhabit. If the work of this book could

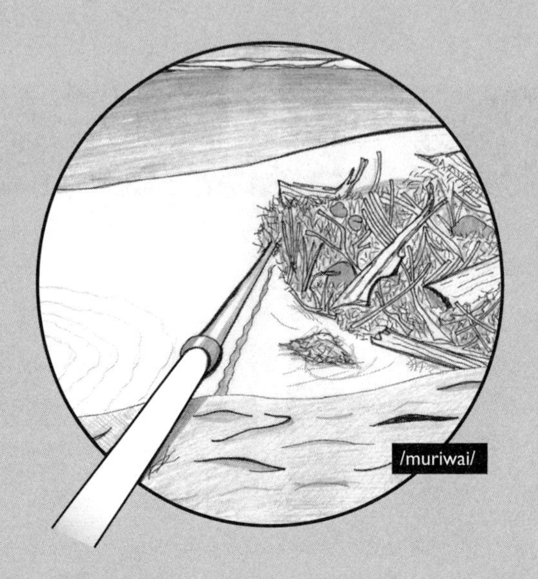

/surfacing/

FIGURE C.1. A cable surfaces at Muriwai Beach, New Zealand.

be captured in a single action, it would be the attempt to grasp a surfacing cable before it recedes, to connect the histories of network infrastructure to today's media environment.

There is a strong consensus in the cable industry that such submerged histories matter. Frank M. Tuttle Jr., the director of overseas engineering and operations at AT&T in the 1980s, commented that "the extent to which each new technology builds on its predecessor is a never-ending revelation," and he argued that the then-emerging fiber-optic networks would be "best understood and managed by those who have a thorough understanding of existing systems."[1] In interviews with me, cable engineers, managers, and even entrepreneurs have echoed this statement. The industry depends on the tried and true, stores much of its knowledge in people, and often establishes new networks along existing routes. This book is an archaeology of undersea networks. It drags up the histories of undersea cables to show how this infrastructure has been produced in and transformed by the environments it extends through; by histories of colonialism, Cold War tensions, and national security; and by the seemingly insignificant actions of residents in coastal communities. These circulations—with their varied scales, geographies, and cultures—constitute a fluid and formative context for today's global networks.

This submarine excavation reveals how the distribution of media and communication does not exist in a naturally smooth space, but requires ongoing

FIGURE C.2. This photograph was printed in a local newspaper with the caption "Debbie Tucker tries her luck at getting a line on the historic undersea telegraph cable uncovered at Mon Repos beach." "Rough Seas, Low Tides Expose Historic Cable," *News-Mail Bundaberg*, April 27, 1990.

spatial manipulation to generate the illusion of friction-free movement. Strategies of insulation and interconnection produce contours that give information traction in the routes it traverses. While strategies of insulation channel signal traffic through environments, strategies of interconnection ground it in local circulations. The creation of stable circuits is a fundamentally environmental process, and as a result, the expansion of the Internet is a distinctly spatial project. Observing the ecologies of communication infrastructures—not how they function as environments, but the ways that they are inextricably embedded in the environments they are designed to transcend—*The Undersea Network* counters the rhetorical pull of terms such as *flow*, which too often connote deterritorialization and dematerialization. This project prompts us to consider the fact that network infrastructure is often not only wired but also centralized, territorial, precarious, and rural. In this conclusion, I outline how this revised conceptualization of cable infrastructure translates into a politics of undersea networks.

Materiality—the physical properties and conditions of a space, including atmospheric, geological, biological, and thermodynamic systems—is often seen as a phenomenon that is overcome and flattened out by communication. In the second chapter of the book, I argued that this perception has been created in part by the representations of cables themselves. Discourses of connection and disruption focus on the cable system prior to its initiation or the moments when it is disconnected, and function as strategies of insulation that keep material environments and operational cables out of view. Even the conventional cable map, depicting the cable as a line between two endpoints and leaving out historical information, portrays the cable as a vector rather than as a material infrastructure. Cable systems have long remained on the periphery of public view, in part because of the limits of existing cable narratives.

Developing a politics of undersea networks depends first on making these infrastructures broadly intelligible. As Lisa Parks argues, when we don't know about the communication infrastructures that support our network society, they tend to remain beyond the limits of public consideration or political engagement.[2] Moreover, as information about their operations and geographies is widely disseminated online and such knowledge becomes a contested terrain, protecting these infrastructures requires the creation of new modes of engagement and new forms of cable visibility. This increasingly has been the position of the cable industry. Lionel Carter, the marine environmental advisor of the International Cable Protection Committee told me that public understanding of the cable network is critical because many of the industry's current problems stem from people's lack of knowledge. He sees education as a way to lower unintended disruptions and facilitate effective policy.[3] Moving away from the limitations of narratives of connection and disruption, chapter 2 highlighted how nodal narratives and narratives of transmission might encourage the exploration of the embodied and environmental dimensions of cable networks, short-circuiting the flow of power via cable discourse.

The other chapters took this approach, tracking undersea cables to reveal the politics of their ownership, regulation, contestation, and development. The book itself is a narrative of transmission, following the signal as it moves through from stations to landings and from islands to oceans and as it stops along the way to detail the histories of individual nodes—tracing how the cultural histories at each point have affected the development of the network as a whole. The first chapter, to counter the technological assumptions that typically govern our understanding of cable networks, zoomed in on the deep social history of transpacific routes. They have been shaped by strategies of insulation devised to protect the network: from the British All-Red Line, which routed cables underwater to avoid land-based turbulence caused by anticolonial uprisings, tariffs, or national conflicts; to the Cold War era, in which many communications hubs were decentralized; and today's commercial practices, which make it difficult to vary where cables are routed. Transpacific routes gain traction in existing contours, leaving us with a global system that is relatively centralized and not very diverse. Rather than see cable networks in relation to the cloud, this chapter asked us to see it as a set of deep channels paid for by hegemonic powers and institutions. We need critical assessments of alternative routes and governmental and public support for cable infrastructure if we wish to have a robust international network or an equitable digital sphere.[4] Otherwise, we might have to plan our information practices for precarious conditions.

The third chapter historicized cable stations—the nodes of our international network—to show how the boundary between the inside and outside of the network has always been negotiated in relation to local publics and environments: it has shifted from the body of the cable worker, to the architecture of the station and the circulation of information about cable operations. Over the years, a cable community has formed to secure and insulate knowledge about the system. Although privatization has created opportunities to open up the production of cable systems to new actors, one of the most vulnerable sites for international cabling is the shortage of skilled workers and the devaluation of their labor. The chapter brought into relief the need not only for insulation, but also for connections that link cable systems to labor and finance for network maintenance, to power systems and electrical grids, and to the ongoing spatial reorganizations of local publics around the cable station.

The fourth chapter looked more extensively at how cables take shape in relation to local ecologies. In the past decade, many studies have begun to address the geographies and materiality of the Internet, but this research tends to focus on urban hubs. Chronicling a set of nodal narratives at the cable landing point, chapter 4 revealed the ties between cable networks and the rural and coastal areas in which cables operate. It documented the cable landing as a pressure point at which cable traffic has been held up by local actors: from the Hawaiian communities whose members contested the running of cables under their roads to the anchors of fishermen. These circulations generate turbulence for cable companies as they seek to traverse the landing point, and in reaction, the companies have designed strategies of insulation to protect cable routes, including the burial of cable underground and the dissemination of cable charts. Ultimately, as in the previous chapters, chapter 4 shows that these strategies give traction to subsequent systems. These forms of turbulence might be reduced by a more cohesive national regulation of the cable landing point. However, we might also imagine strategies of interconnection that ground networks in local circulations rather than buffering out those circulations, and that give local actors an interest in the cable's continued operation.

Existing strategies at the cable landing point can benefit local communities whose members gain power via local or state bureaucracy, but at the same time these strategies can make it difficult for other countries to connect with the Internet. Chapter 5 pushed back against the dominant assumption that islands are sites off the network by sketching an archipelago of nodal narratives. It looked at networked islands like Guam and Fiji as critical nodes: places where transpacific information exchange—often seen as virtual lines of connection—can be understood as a material, geographic, and environmental process and where existing currents can be grounded and triangulated. The chapter also revealed the support needed by many islands in order to interconnect with cable networks. Not every location in the world will get cables, even leftover ones, given the current economic model. If we truly value democratizing communication infrastructure, governments or other organizations will have no option but to subsidize these systems. The chapter also shifted our understanding of progress in content production: without the extension of equitable infrastructure networks, the transition to high-bandwidth content may lead to further inequality.

Undersea cables' movement through distant aquatic environments makes it possible for them to interconnect with other oceanic organizations and actors. Chapter 6 showed how cable systems are shaped in relation not only to anticipated interference imagined in global geopolitics, but also to the materiality of aquatic spaces. Tracing the residual uses and knowledge effects of undersea cables, it sketched out how these networks might be leveraged during or after their commercial lifespan: they can be used for activities from military monitoring to oil extraction and global climate science. Developing alternative uses (and users) of these systems is one way to ensure the broader equity of communications systems across the Pacific.

Considering undersea cables as an environmental technology, we might look to the ways that they have facilitated the coordination of movement through other frontiers. The Commonwealth Pacific Cable (COMPAC) and South-East Asia Communications (SEACOM) coaxial cables were critical for the National Aeronautics and Space Administration's early space exploration; and they were key to various U.S. missions and defense operations, including the launching of the first American into space.[5] An ad for Standard Telephone and Cables (STC) claims that the "demands of space exploration have taken STC to the bottom of the ocean."[6] Another transoceanic cable route in the 1960s (which never materialized) was planned to link four sea stations outfitted with radio communications in a corridor across the North Atlantic and would have helped make transatlantic air transport, which then depended on VHF radio, more reliable.[7] Each sea station would have had five decks; extended eighty-three feet up out of the water; and carried a crew of sixteen men, as well as meterological and oceanographic instruments for reporting the weather and studying the oceans. The U.S. Underseas Cable Corporation reported on their potential strategic function: "Seastations really provide 'instant real estate'—islands that can be put practically anywhere in the world's ice-free oceans, islands that could serve the great missile test ranges, as tracking, telemetry, and control stations."[8] Undersea cables not only facilitated the production of knowledge about the depths, but they have also produced other wired ecologies, including those in the extraterritorial skies and outer space.

In all of these environments, global signal traffic carried by undersea cables must be insulated from turbulent ecologies by strategies of spatial manipulation. Through this manipulation, currents gain traction in particular topographies, which we can excavate by diving down through their layers. An archaeology of these systems challenges the use of fluidity as the metaphor for movement in contemporary information environments. Zygmunt Bauman describes our social sphere as a liquid modernity, where "for power to be free to flow, the world must be free of fences, barriers, fortified borders and checkpoints. Any dense and tight network of social bonds, and particularly a territorially rooted tight network, is an obstacle to be cleared out of the way. Global powers are bent on dismantling such networks for the sake of their continuous and growing fluidity, that principle source of their strength and the warrant of their invincibility."[9] The apparent liquidity of social processes is certainly facilitated by the undersea cables that carry over 99 percent of data traffic between continents: mobility is predicated on these relatively static infrastructures. However, the fluidity of global information systems leads to the breaking down of some spatial barriers but is predicated on the construction of elaborate strategies of insulation, fixed structures that shelter power from natural and social threats, and strategies of interconnection that makes it preferable to link systems with existing networks rather than branch out to new sites. Since engineers have often been in charge of organizing cabled spaces, it is not surprising that logics fundamental to electrical engineering—insulation and interconnection—dominate the network's cultural geography. Fluidity is not a process that simply overcomes fixed structures; it also requires them. Our seemingly wireless lives are predicated on a mess of tangled wires.

This book has presented a relationship between fixity and fluidity similar to that described by Lieven de Cauter, who observes that the network society is a space for alliances and connections and simultaneously "a society of exclusion and seclusion. . . . The image of the society of the future is perhaps that of the entropic universe: islands of order and increasing complexity in an ocean of chaos."[10] For de Cauter, this process—which he terms *encapsulation*—means that the increase in boundary crossing is accompanied by a less often observed investment in borders and protective enclaves. If the capsule increases human mobility, insulation is what shelters the transmission of global information flows from local ecologies. The faster and more reliable global signal traffic becomes, the more insulation these signals will require. It is the process of insulation that has facilitated the sedimentation of cable routes. The future development of the Internet, rather than decreasing the significance of space, will depend on ever more extensive investments in cabled sites. Even though our media environment appears to be speeding up, we must also pay attention to the fixed material investments that ground today's networks.

This book has examined only one part of our global Internet infrastructure—the segment that runs under the oceans—but the historical research and analytic concepts presented here can help us understand the environmental dynamics of other distribution infrastructures, from the data center to the Internet exchange. It is important to understand not only the nebulous and cloudlike dimensions of these systems, but also the development of their material foundations and the points at which they intersect with (and affect) existing ecologies—especially since as the Internet expands, more systems will be established, more energy will be needed to keep facilities cool, and technologies will be expanded to maintain them. Moreover, the concepts here can be extended to characterize the ecological dimensions of media more generally. What strategies of interconnection have been developed to facilitate the transduction of signals between different kinds of media environments? What strategies of insulation have been devised to keep the circulation of films and other materials from diffusing into a media environment that is too conductive? How have these given certain objects, texts, or phenomena traction in some mediascapes and not in others? These questions are increasingly pertinent in an information ecology where the problem is not how to overcome fixed barriers, but how to navigate in a sea of too much information. We might reconceptualize all media today as being under water, a starting point that compels us to examine how they avoid, are insulated from, and are anchored in existing sets of circulations.

If we think of the Internet as only a virtual environment, then our conception of how to change it will depend on changing only the virtual world. This study of undersea cables, the material infrastructure of the Internet, has led in contrast to a set of tangible policies and politics. These include the diversification of cable narratives and routings, as well as the development of media content for variable levels of access and precariousness. It entails revaluing cable labor and developing a more extensive cable workforce, interventions into the bureaucracy of the landing point, and the better engagement of local communities. It involves a call for state support for new cable networks, the triangulation of currents between existing nodes, and the expansion of networks' potential uses. As the undersea cable network currently exists, hooking into it often increases one's dependence on a relatively narrow range of existing actors, powers, and routes. Together these politics and policies would instead give traction to more broadly robust, resilient, and equitable global networks.

NOTES

introduction

1. Rauscher, *ROGUCCI Study Final Report*, 9.

2. Quoted in Rauscher, *ROGUCCI Study Final Report*, 9.

3. Rauscher, *ROGUCCI Study Final Report*, 38.

4. The number of international undersea cable systems was calculated based on TeleGeography's *Submarine Cable Landing Directory*, accessed January 2014, http://www.telegeography.com/telecom-resources/submarine-cable-landing-directory/index.html.

5. Liu and Pound, "The Amoderns."

6. Rauscher, *ROGUCCI Study Final Report*, 131.

7. Star, "The Ethnography of Infrastructure," 380.

8. Interview with Stewart Ash, marine design and installation manager, WFN Strategies, January 20, 2012.

9. Mackenzie, *Wirelessness*.

10. Interview with anonymous cable station manager, May 24, 2011.

11. Steinberg, *The Social Construction of the Ocean*.

12. M. Taylor, *Confidence Games*, 154.

13. See, for example, Cookson, *The Cable*; Headrick, *The Tools of Empire* and *The Tentacles of Progress*; Kennedy, "Imperial Cable Communications and Strategy"; Winseck and Pike, *Communication and Empire*.

14. Parks, *Cultures in Orbit*.

15. For an overview of the competition between the two technologies, see Warf, "International Competition between Satellite and Fiber Optic Carriers."

16. *Optical Fibres versus Satellites*, exhibition in Porthcurno Telegraph Museum, Porthcurno, United Kingdom, July 2011.

17. Mackenzie, *Wirelessness*, 64–65.

18. Blum, *Tubes*, 5.

19. A danger in writing about critical infrastructure is that the information collected could in theory be used to disrupt these networks. However, much of the supposedly sensitive information I refer to is already available online.

20. Interview with John Hibbard, president of the Pacific Telecommunications Council, April 27, 2011.

21. Findley, "Now You See It."

22. Interview with Fiona Beck, CEO of Southern Cross Cable Network, January 16, 2012.

23. Organizations such as the International Cable Protection Committee have sought to strategically increase the visibility of these systems.

24. Anna Munster has argued in her book on this topic that "it is the map of distributed communications that has become the definite image of technically inflected networks of communication" (*An Aesthesia of Networks*, 23).

25. Galloway, "Networks," 290.

26. Calculated in January 2014, using data from TeleGeography. Cables included were operational at that time and connected to locations outside the United States.

27. PIPE Networks, "Transmission Technology (Part 2)."

28. Baran, "On Distributed Communications."

29. P. Edwards, *The Closed World*; Vaidhyanathan, *The Googlization of Everything.*

30. Bernard Finn observes that the basic technical elements of undersea telegraphy "were established early and remained largely unchanged for a century" ("Submarine Telegraphy," 11). Jean-François Blanchette has made a similar observation about computing history: "the solutions developed in a particular moment of technical history tend to persist and accrete. Thus, in contrast to the perception of computing as moving at a furious pace of technical evolution, its infrastructure evolves very slowly" ("A Material History of Bits," 1054).

31. See Elsaesser, "'Constructive Instability'"; Sampson, "The Accidental Topology of Digital Culture."

32. See Ford-Ramsden and Burnett, "Submarine Cable Repair and Maintenance."

33. Rauscher, *ROGUCCI Study Final Report*, 11.

34. Graham and Thrift, "Out of Order," 7.

35. See Curtin, "Media Capitals"; Graham and Marvin, *Splintering Urbanism*; Graham, *The Cybercities Reader*; Larkin, *Signal and Noise*; Sassen, *The Global City*; Dodge and Kitchin, *Mapping Cyberspace*; Malecki, "The Economic Geography of the Internet's Infrastructure."

36. Kirschenbaum, *Mechanisms*; Bozak, *The Cinematic Footprint*; Cubbitt, Hassan, and Volkmer, "Does Cloud Computing Have a Silver Lining?"; Fuller, *Media Ecologies.*

37. Parks, *Cultures in Orbit*; Dourish and Bell, *Divining a Digital Future*; Downey, "Virtual Webs, Physical Technologies, and Hidden Workers"; Graham and Marvin, *Splintering Urbanism*; Larkin, *Signal and Noise*; Jackson, Edwards, Bowker, and Knobel, "Understanding Infrastructure"; Edwards, Jackson, Bowker, and Knobel, "Understanding Infrastructure."

38. Chun, *Control and Freedom*; Lovink, *Dark Fiber.*

39. Dourish and Bell, "The Infrastructure of Experience and the Experience of Infrastructure," 424.

40. Starosielski, Soderman, and cheek, "Introduction"; Liu, "Remembering Networks."

41. Cunningham, "Can the Arctic Provide an Alternative Route?"

42. The Polarnet project recently announced an $860 million system from the United

Kingdom through the Siberian Sea to Vladivostok and on to Tokyo (D. Scott, "Dive, Dive and I'll Connect You to Siberia").

43. Tsing, *Friction*.

44. Moin and Kim, "Tackling Turbulence with Supercomputers," 62.

45. De Certeau, *The Practice of Everyday Life*, 36.

46. Cunningham, "Can the Arctic Provide an Alternative Route?" 11.

47. Brewin, "Northwest Passage Fiber Optic Line Could Support Defense Arctic Strategy."

48. Brewin, "Northwest Passage Fiber Optic Line Could Support Defense Arctic Strategy," 5. Indeed, a planned Arctic cable by the Kodiak-Kenai Cable Company fell through only a few years earlier, after failing to secure stimulus funding from the U.S. government (Torrieri, "Kodiak-Kenai Cable Company Kicks Off Undersea Arctic Fiber Optic Project").

49. For example, as Manuel Castells notes, developing interconnections between systems is a primary source of power in the network society. For him, the connecting points are "switches" ("A Network Theory of Power"). In science and technology studies, these strategic intermediaries are termed gateways. Jackson, Edwards, Bowker, and Knobel, "Understanding Infrastructure."

50. Sterne, *The Audible Past*; Helmreich, *Alien Ocean*.

51. Rauscher, *ROGUCCI Study Final Report*, 60.

52. Djelic and Quack, "Overcoming Path Dependency."

53. Morley, "Communications and Transport," 756.

54. Neil Postman, "The Reformed English Curriculum"; Fuller, *Media Ecologies*. Here, following Jussi Parikka ("Introduction"), I see media ecology as the description of how media exist in and as part of ecology.

55. In this approach, I've drawn on multisited ethnography, a method that attempts to grasp large-scale world systems by following objects in circulation and documenting the worlds in which they materialize. See Marcus, "Ethnography in/of the World System."

56. These included the Micronesian Area Research Center in Guam, the Hawai'i State Archives, the Hawaiian and Pacific Collections at the University of Hawai'i at Manoa, the Bishop Museum, the Environmental Design Archives and the Bancroft Library at the University of California at Berkeley, the Fiji National Archives, the University of the South Pacific's Pacific Collection, Archives New Zealand, the Alexander Turnbull Library, the Auckland Museum, the Auckland Museum of Transport and Technology, the Auckland Library, the Takapuna Library, the Radio New Zealand Sound Archives, the Far North Regional Library, the Northern Territory Library, the National Archives of Australia in Canberra and Brisbane, and the company archives of Cable & Wireless at Porthcurno in the United Kingdom and AT&T in Warren, New Jersey.

ONE. circuitous routes

1. Rauscher, *ROGUCCI Study Final Report*, 24.

2. Rauscher, *ROGUCCI Study Final Report*, 24.

3. Main, "The Global Information Infrastructure," 88.

4. For example, Junho H. Choi, George A. Barnett, and Bum-So Chon argue that the "most appropriate geographic units for the analysis of telecommunications and transportation networks are the metropolitan areas that play key roles in global interactions" ("Comparing World-City Networks," 82).

5. Cunningham, "Can the Arctic Provide an Alternative Route?"

6. Graham and Marvin, *Splintering Urbanism*, 30.

7. Hills, *Telecommunications and Empire* and *The Struggle for Control of Global Communication*; Winseck and Pike, *Communication and Empire*.

8. Kelty, "Against Networks," 5. See also T. Hughes, *Networks of Power*, introduction.

9. Warf, "International Competition between Satellite and Fiber Optic Carriers," 5.

10. The location was also selected because of the security afforded by military and defense installations. P., "A Short Account of La Perouse."

11. In his analysis of how Valentia Island was chosen for the landing of the Atlantic Cable, Donard de Cogan suggests that justifications for landings "were a confused mixture of science and self-interest": in this case, Sir Patrick Fitzgerald, a well-connected resident landlord, mobilized the shortest-distance argument, "lavishly" entertaining those connected with the cable project ("Background to the 1858 Telegraph Cable," 23). At the other end of the Atlantic, more than one history has hinted that the choice of the landing point at Heart's Content might have something to do with the name of the village (Rowe, *Connecting the Continents*).

12. Willoughby Smith, *The Rise and Extension of Submarine Telegraphy*, 11.

13. The need for gutta-percha led to enormous environmental devastation on the Malay Peninsula as trees were chopped down to provide for cable insulation. See Headrick, "Gutta-Percha"; Picker, "Atlantic Cable."

14. Earlier in the century, supporters of the semaphore system had pointed out the vulnerability of wires to attack and the relative reliability of the optical telegraph. See Parikka, "Mapping Noise," 265.

15. Quoted in Headrick, "Strategic and Military Aspects of Submarine Telegraph Cables 1851–1945," 189.

16. Pacific Cable Act 1901, *Account 1907–1908*, 7.

17. V. Hughes, "Cable Connections."

18. Nalbach, "The Software of Empire," 76.

19. Knuesel, "British Diplomacy and the Telegraph in Nineteenth-Century China"; Ahvenainen, *The Far Eastern Telegraphs*.

20. Headrick, *The Tentacles of Progress*, 107.

21. Smithies, "The Trans-Tasman Cable," 701.

22. Winseck and Pike, *Communication and Empire*.

23. See Müeller-Pohl, "Colonialism, Decolonization and the Global Media System"; Winseck and Pike, "The Global Media and the Empire of Liberal Internationalism," 34.

24. Müeller-Pohl, "Colonialism, Decolonization and the Global Media System."

25. Müeller-Pohl, "Colonialism, Decolonization and the Global Media System."

26. The company originally considered landing on Hawai'i's Necker Island (as well as the Gilbert Islands and the Solomon Islands), which was over $2 million less expensive and better interconnected with existing transport networks but ultimately decided on

Fanning because it was felt that security would be compromised otherwise. The company chose not to connect with a French cable between New Caledonia and Australia after the Australian press launched a campaign to support an all-British route. See Finn, "Introduction."

27. The Pacific Cable Board had paid £200 for each call to the island. It had hoped that companies would commercially exploit the islands (and further justify trade), but this never happened. "Telegraph Cables (Further Papers Relating to), No. 13," Papers Presented to both Houses of the General Assembly by Command of His Excellency, *New Zealand*, 1914, New Zealand National Archives, Wellington, New Zealand.

28. Müeller-Pohl, "Colonialism, Decolonization and the Global Media System."

29. Müeller-Pohl, "Colonialism, Decolonization and the Global Media System."

30. Winseck and Pike, "The Global Media and the Empire of Liberal Internationalism," 34.

31. Winseck and Pike, "The Global Media and the Empire of Liberal Internationalism," 34.

32. The Pacific Cable Act of 1911 later authorized the construction of duplicate parts of the transpacific cable for diversity. ("Telegraph Cables [Further Papers Relating to], No. 13"). The Eastern Telegraph Company reported in 1914–15 that since it aimed to make its service "as secure as possible against total interruption," it had taken the "precaution of connecting the most important point of [the] system by several cables laid along widely different routes" ("The Cable Companies and the War," 236).

33. Kieve, *The Electric Telegraph*, 111.

34. As Bernard Finn notes in his study of undersea telegraphy, the design of the cable and techniques for manufacture and installation also remained almost unchanged for nearly a century, a factor that further reinforced the industry's tendency to stick to the tried and true ("Introduction").

35. "Nerves of the Navy," 270.

36. Quoted in "Commentary on a Premature Burial (at Sea)."

37. Schwoch, *Global TV*, 5.

38. Interview with anonymous cable engineer, October 29, 2010.

39. Finn, "Introduction." Some wondered if undersea telegraphs would ever disappear. One writer speculated: "When will the last ocean cable direct-current telegraph service be shut down? Perhaps about the time the last railroad train is retired" ("Commentary on a Premature Burial [at Sea]").

40. For example, the 1976 cable from Auckland to Sydney was seen as offering "Trans-Tasman diversity" that would help the nations to avoid "dependence on the satellite system" and "reduce the effects of system failure" (G. Press and Stringfellow, "The Second Trans-Tasman Cable System," 333).

41. Vanderbilt, *Survival City*.

42. For example, the station at La Perouse, Australia, was moved to central Sydney, and the cable laid to New Zealand's South Island was relocated to the capital, Wellington. Many remote stations were closed, including Broome and Darwin.

43. Its distance from a port was one reason that Bamfield, Canada, was reportedly chosen as a location for the Pacific Cable. In the 1920s, due to the increased disrup-

tions caused by shipping at the port of Penang in Southeast Asia, the Eastern Telegraph Company shifted six cables to a landing point miles outside the harbor ("Ships' Anchors Interrupt Shore-Ends of Cables at Penang and Cause Landing Place to be Shifted Six Miles").

44. The first telephone cable from the United States was laid to the east side of O'ahu in 1957; subsequent cables were landed on O'ahu's west shore.

45. "Possible Sites for Cable Landing and Shore Installations in Australia, Pacific Cable Conference 1959," September 1959, 1–2, PCC 59/16, National Archives of Australia, Canberra, Australia.

46. Interview with anonymous cable engineer, October 29, 2010.

47. The SEACOM cable was an exception to this. In SEACOM's case, the government decided that the cable should exit out of Cairns to justify the creation of a national backbone, a decision that later made the network unreliable. As one engineer recalls, SEACOM's microwave link "was nothing but trouble. The cable would often just continually fail. Not because of the cable system, but the backhaul link between Cairns and Sydney" (interview with Geoff Parr, global manager, Transmission Network Engineering & Network Buildings, May 24, 2011).

48. "Pacific Cable Conference 1959 Vol 1, Report to the Governments of the United Kingdom, Canada, Australia, and New Zealand. Sydney, October 20, 1959," p. 19, Pacific Cable Conference—Documents, 1838, 707/49 ANNEX, National Archives of Australia, Canberra, Australia.

49. For ANZCAN, which replaced COMPAC, engineers considered several other landings but decided to follow the existing route since it had not had a single break due to seabed conditions—to diverge from what had been proven secure was to take an unnecessary risk.

50. Rauscher, *ROGUCCI Study Final Report*, 79.

51. The U.S. Underseas Cable Corporation also fulfilled nonmilitary orders, including repeaters for the Japan-Siberia cable in 1969. At times the company's ships "stood cable guard" over its systems ("Neptune," 5).

52. As one cable management committee reported, "AT&T would probably be compelled for military reasons to provide a Guam/Manila telephone cable although they might not have been inclined to do so on purely commercial grounds." ("Commonwealth Cable Management Committee: First Meeting, Sydney 24th–26th November, 1965," PCMC 1963/8, Archives New Zealand, Wellington, New Zealand).

53. The year before, the Japanese parliament had created the Nippon Telegraph and Telephone Public Corporation as a monopoly domestic provider of communications.

54. S. Shimura, *International Submarine Cable Systems*, 11.

55. The major players were British Telecom and Standard Telephone Cable in Britain and France Telecom and Alcatel in France (Devos, "Changing Relationships and Responsibilities").

56. "Pacific Cable Conference."

57. Collins, "Working Paper," 4.

58. Devos, "Changing Relationships and Responsibilities," 211.

59. "Pacific Cable Conference." Instead AT&T purchased capacity on COMPAC, though

the company's traffic on these lines had to terminate in the United States. It could connect to other systems only if it was not aiming to compete for national traffic.

60. Although having many entities cofinance the construction would facilitate international cooperation and funding, with too many parties there would be divergences of opinion and delays in system development. This approach to cable laying also represented a divergence from the organization of satellite communications, in which control was spread among more countries. With a cable system, there tended to be a small number of owners and "a common purpose" (interview with Gregory Sachs, cable engineer, May 23, 2011).

61. Devos, "Changing Relationships and Responsibilities," 211.

62. Interview with anonymous cable engineer, January 20, 2012.

63. Interview with anonymous cable engineer, October 29, 2010.

64. "Pacific Cable Conference," 13.

65. Expenses for the cable were predicted to be over £2,500,000 a year ("Pacific Cable Conference," 13).

66. "Pacific Cable Conference," 19.

67. This connection extended the U.S. Defense Communications System through Australia, which required five circuits to Sydney for the first ten years of the cable's life. See "Cables in the Pacific"; "Commonwealth Cable Management Committee," Archives New Zealand.

68. "License Authorizing the Landing and Operation of Two Submarine Cables on the Island of Oahu, Hawaii, FCC 61–258, 99661, File No. S-C-L-20," Trans Pacific Cable Conference and South East Asia Conference, A1838, 707/49 part 3, National Archives of Australia, Canberra, Australia.

69. Interview with Parr, May 24, 2011.

70. Via Guam, these were interconnected in an East Asian loop. GPT extended from Tanguisson Point to Infanta (Philippines) and Fangshan (Taiwan). From Fangshan, HON-TAI ran to Cape D'Aguilar, Hong Kong, where HJK linked to back to Chikura, Japan, and Cheju, Korea.

71. Local knowledge sometimes played a role in determining the safest routes. In the case of one network, local residents had information about places where Americans had dumped ammunition, prompting cable companies to conduct more detailed underwater surveys and a mine clearance operation. In another case, on the eastern coast of Australia, engineers thought that their cable had reliably crossed a creek, but was taken out by a storm. A local resident later gave them information about the storms in the area that helped the company to determine an alternative, safer route.

72. Interview with Anthony Briscoe, general manager, international, Telecom New Zealand, October 14, 2010.

73. Costin, "The Impact of Australian Telecommunications Deregulation upon International Cable Planning."

74. This included a second generation of optical amplifier systems and a process called dense wavelength division multiplexing. Although first-generation fiber technologies regenerated fading signals on the seafloor (a process that limited the cable's capacity to the amount available during the design period), new technologies meant that

companies could simply put new terminal equipment on either end and increase the capacity (Warf, "International Competition between Satellite and Fiber Optic Carriers," 6).

75. Warf, "Reach Out and Touch Someone," 257.

76. Reinaudo, "New Market Calls for New Business Practices," 15.

77. Reinaudo, "New Market Calls for New Business Practices," 14.

78. Interview with anonymous cable entrepreneur, October 2010.

79. This diverged from the historical practice of forcing companies to buy fixed amounts at the outset. See Beaufils and Chabert, "Interconnection of Future Submarine Cables in the Global Communications Web," 94.

80. Interview with anonymous cable operations manager, May 26, 2011.

81. Interview with Stewart Ash, associate, WFN Strategies, January 20, 2012.

82. Edward Malecki and Hu Wei observe that the number of countries linked by undersea cables tripled between 1979 and 2005 ("A Wired World," 368).

83. These players included AT&T Submarine Systems and Simplex (later TE SubCom) and Simplex (later SubCom), Alcatel-Lucent (which merged with the British Standard Telephone and Cables Submarine Systems in 1994), and the Japanese companies KDD Submarine Cable Systems and NEC (which owned Ocean Cable Company). Prior to this shift, national cable operators each had their own cable ships. In a competitive environment, telecommunications companies could no longer afford the cost of marine maintenance infrastructure and sold it off. The marine segment continued to be an industrial pressure point, with supply (and access to cable repeaters) consolidated among these few players. See Devos, "Changing Relationships and Responsibilities."

84. Ruddy, "An Overview of International Submarine Cable Markets."

85. For a week-long cable break, the revenue lost could be greater than 1 percent of the cable's value (Reinaudo, "New Market Calls for New Business Practices," 14).

86. Interview with Brett O'Riley, telecommunications policy expert, October 1, 2010.

87. These networks were the China-U.S. network (which had nine landings in five countries in 1999), Pacific Crossing (1999), Japan-U.S. network (2001), and Tyco Global Network Pacific (2002).

88. Gonze, "U.S./Japan Trade War Broils."

89. KDD, France Telecom, and Singtel (now private rather than government backed) laid the SEA-ME-WE 3 along the same route as FLAG. Following NPC's development, AT&T, KDD, and Teleglobe Canada laid Trans-Pacific Cable 4 across the North Pacific.

90. Interview with anonymous telecommunications worker, October 2010.

91. Interview with Charles Jarvie, head of technology and operations, Telecom New Zealand, September 22, 2010.

92. Interview with Brett O'Riley, October 1, 2010.

93. Interview with anonymous telecommunications worker, October 14, 2010.

94. Interview with Fiona Beck, CEO of SCCN, January 16, 2012.

95. Interview with anonymous telecommunciations worker, September 22, 2010.

96. Warf, "International Competition between Satellite and Fiber Optic Carriers," 8.

97. Warf, "International Competition between Satellite and Fiber Optic Carriers," 7.

98. Interview with anonymous cable engineer, January 20, 2012.

99. In this decade, much of the capacity was still primarily devoted to transmitting voice traffic. See Dadouris, Singhi, and Long, "The Impact of the Internet and Broadband Service Offerings on Submarine System Cable Capacity into the 21st Century."

100. Ruddy, "An Overview of International Submarine Cable Markets."

101. "Breaking News."

102. Warf, "Reach Out and Touch Someone," 257.

103. Interview with Stewart Ash, January 20, 2012.

104. Kedar and Duplantie, "Building the International Infrastructure for the Second Tier Market," 55.

105. Esselaar, Gillwald, and Sutherland, "The Regulation of Undersea Cables and Landing Stations."

106. *Cables and the Earth*, exhibition in Porthcurno Telegraph Museum, Porthcurno, United Kingdom, July 2011.

107. Interview with anonymous cable station manager, May 24, 2011.

108. Even if the ship comes from another zone, it might suddenly be needed back there and have to drop the cable it is working on to return.

109. Watts, "Maritime Critical Infrastructure Protection."

110. Rauscher, *ROGUCCI Study Final Report*, 77; Bressie and Findley, "Cybersecurity Developments Raise Growing Regulatory Concerns for Undersea Cable Industry."

111. Tagare, "Risk Management Is Key to Submarine Cable Success."

112. At least since the 1960s, the Department of Defense has used commercial cables, benefiting from system bandwidth improvements, and has been interested in the security of cable routing. One engineer told me that the department has an interest in cable circuits that land only on U.S. territory, without stopping at other sovereign nations (interview with anonymous telecommunications worker, 2011).

113. Interview with anonymous cable manager, November 2010.

114. Interview with John Humphrey, Pacific Fibre, October 25, 2010.

115. Interview with John Hibbard, April 4, 2011.

116. Graham and Thrift, "Out of Order," 5.

117. Rauscher, *ROGUCCI Study Final Report*, 84. After the Fukushima disaster, for example, cable builders may think twice about landing cables near a nuclear power plant.

118. From 1876 to 1902, the Eastern Telegraph Company monopolized telegraph traffic into and out of the country and could set prices as high as it liked. It was not until the Pacific Cable Board was laid from Canada in 1902 that competition started.

119. Up to 90 percent of New Zealand's Internet traffic is destined for the United States. See "Entrepreneurs Announce $900m Fast Broadband Plan."

120. It was backed by the Internet entrepreneur and multimillionaire Sam Morgan, inventor of TradeMe, the New Zealand version of eBay. See Hendery, "Strength in Telco Networking Numbers."

121. Several companies have attempted to lay a second cable to New Zealand. Kordia, a government-owned company, announced plans to build OptiKor from Australia in 2008, only to shelve the project three years later after failing to obtain funding. In 2011 the project was taken over by the Chinese companies Axin and Huawei Marine. See Fletcher, "Kordia Puts Brakes on Cable."

122. Doesburg, "Across the Tasman, Two Cables Better Than One."

123. Interview with anonymous cable entrepreneur, October 2010.

124. Interview with anonymous cable entrepreneur, October 2010.

125. Interview with anonymous cable entrepreneur, October 2010.

126. Interview with Anthony Briscoe, general manager, international, Telecom New Zealand, October 14, 2010.

127. Dellinger, "The Million Dollar Millisecond."

128. Interview with John Humphrey, October 25, 2010.

129. They have also suggested that the cable could make New Zealand a financial hub. "Entrepreneurs Announce $900m Fast Broadband Plan."

130. Interview with John Humphrey, October 25, 2010.

131. The landing might even benefit the network by providing a backup power source.

132. It is often difficult to get development funding from institutions such as the Asian Development Bank or the United Nations. Interview with anonymous cable entrepreneur, October 2010.

133. Interview with anonymous cable entrepreneur, October 2010.

134. Interview with Charles Jarvie, head of technology and operations, Telecom New Zealand, September 22, 2010.

135. Interview with anonymous cable entrepreneur, October 2010.

136. Interview with anonymous cable entrepreneur, October 2010.

137. Rauscher, *ROGUCCI Study Final Report*, 93.

138. Interview with anonymous telecommunications worker, September 22, 2010.

139. Interview with anonymous telecommunications worker, October 2010.

140. Fletcher, "New Pacific Cable Link Plan Unveiled."

141. Schmidt, Bradsher, and Hauser. "U.S. Panel Cites Risks in Chinese Equipment."

142. Hedges, "Hibernia Networks Reveals New Partner for Project Express Cable Deployment."

143. Rauscher, *ROGUCCI Study Final Report*, 88.

144. Rauscher, *ROGUCCI Study Final Report*, 19 and 34.

TWO. short-circuiting discursive infrastructure

1. Hayes, "Reminiscences of TAT-1."

2. Parks, "Around the Antenna Tree."

3. Reading the balance of power levels has long been critical to determining the location of faults.

4. Rauscher, *ROGUCCI Study Final Report*, 63.

5. Narratives about the first attempts to establish a transatlantic telegraph helped subsequent projects to gain funding. Donard de Cogan has suggested that Cyrus Field's "surprising" ability to convince a conservative U.S. Treasury to support the Atlantic Cable may have been in part due to his brother's best-selling book about the cable laying ("Background to the 1858 Telegraph Cable," 22). Colin Hempstead similarly argues that narratives about the cable written by the expedition's journalist helped sell British shares in future developments ("Representations of Transatlantic Telegraphy"). Connec-

tion narratives were used, especially in the telegraph era, to educate citizens about communications and stimulate the use of new technologies.

6. Committee members were given authority to carry out publicity for their own country, reach out to local communities, and take advantage of the cable system to benefit prospective customers. See Pacific Cable Management Committee, "Report on Public Relations Meetings, First Series," February 1962, DOC/PCB/1/1 1962, Cable & Wireless Archives, Porthcurno, UK.

7. Their specific objectives included traffic raising, to make sure that traffic would be maximized as soon as possible; prestige, to promote national communications companies; and build-up for the Commonwealth system, to ensure that the Commonwealth cable system would be "universally regarded as the world's most efficient telecommunications system" (Pacific Cable Management Committee, "Report on Public Relations Meetings, First Series"). See also "Schools 'Adopt' the Company's Cableships."

8. Pacific Cable Management Committee, "Report on Public Relations Meetings, First Series."

9. This last film conformed less than the others to a standard narrative of connection, as it tracked the cable for twenty-four hours in its life. However, it was still embedded in a publicity campaign focusing on the cable's connection.

10. "*Call the World* to Be Shown at Two Major Film Festivals."

11. Pacific Cable Management Committee, "Report on Public Relations Meetings, First Series."

12. Pacific Cable Management Committee, "Report on Public Relations Meetings, First Series," 12.

13. "COMPAC—5th Meeting PCMC," p. 24, Record Number 43/7/2, AAMP W3174 909 107, Archives New Zealand, Wellington, New Zealand.

14. Pacific Cable Management Committee, "Report on Publicity, Third Meeting," Record Number PCMC/1961/8, Archives New Zealand, Wellington New Zealand.

15. Pacific Cable Management Committee, "Report on Public Relations Meetings, First Series," 16.

16. See *Bridge under the Ocean* (United States, 1957), *Telephone Cable to Cuba* (United States, 1950), and *Link to the North* (United States, 1957). Cable films, including *Voices from the Deep* (Wallace Bennett, United States, 1969), *A Chord of Voices* (1976), and *Lightwave Undersea Cable System* (United States, 1983), continued to be produced through the next several decades, though with less frequency.

17. The Atlantic Cable also surfaced in the books by L. J. Davis, *Fleet Fire: Thomas Edison and the Pioneers of the Electric Revolution* (2003) and Standage, *The Victorian Internet*; the television documentary *Last Chance Trans-Atlantic* (Germany, 2002); and Barbara Murray's novel *Gifts and Bones* (2006). New research findings and resources were also released during this period, including Gillian Cookson, *The Cable: The Wire That Changed the World* (2003), museum exhibitions such as the Smithsonian's *The Underwater Web: Cabling the Seas* (2001–2), and the popular historical website Atlantic-Cable .com.

18. The Atlantic Cable history has routinely surfaced to help orient us when communications technology is in transition. A cycle of Atlantic cable histories was published

concurrently with the development of undersea telephone cables in the late 1950s. Two of the most often cited are Clarke, *Voice across the Sea*, and Dibner, *The Atlantic Cable*.

19. Gordon, *A Thread across the Ocean*, 215.

20. Warf, "International Competition between Satellite and Fiber Optic Carriers."

21. This trend has precedent in stories that directly parallel Field's biography with the adventure of the cable. See McDonald, *A Saga of the Seas*; Carter, *Cyrus Field*.

22. Standage, *The Victorian Internet*, 75. This observation—that had Field known anything about cable laying, he likely would not have pursued it—is also made in *The Great Transatlantic Cable* and *Transatlantic Cable*.

23. There are other examples of this: The *Great Eastern*, the giant cable ship that laid the 1866 cable, was the cover image for a calendar distributed at the 2009 Pacific Telecommunications Council's conference. The fiber-optic cable consulting agency T. Soja and Associates posted online a series of presentation slides that featured the Atlantic Cable. The industry magazine, *Submarine Telecoms Forum*, has also included historical reflections on the system.

24. Griesemer, *Signal & Noise*, 338.

25. Griesemer, *Signal & Noise*, 289.

26. W. Russell, *The Atlantic Telegraph*, 57.

27. Less than 10 percent of faults are the result of hardware failures (Rauscher, *ROGUCCI Study Final Report*, 79).

28. Singtel, "Cable Cut Fever Grips the Web."

29. Kipling, "The Deep-Sea Cables," 181.

30. Quoted in Dibner, *The Atlantic Cable*, 20.

31. For example, an illustration of the *Agamemnon*'s 1858 laying of undersea cable depicts a large whale chasing the cable-laying ship. A description of the event reads: "A very large whale was seen approaching the cable 'at a great speed.' It appeared to be making straight for the cable and it was a great relief when it was seen to pass astern, just grazing the cable where it entered the water." Attacks from "man-eating" sharks and giant octopi were also highlighted in stories circulated among cable men (Kieve, *The Electric Telegraph*, 98).

32. Hill, "An Underwater Cable and Other Communication News and Views."

33. Hill, "An Underwater Cable and Other Communication News and Views."

34. J. Edwards, "Deep Sea Interlude," 367.

35. Clarke, *Voice across the Sea*, 1.

36. Schwoch, *Global TV*.

37. See Heezen, "Whales Entangled in Deep Sea Cables"; Shapiro et al., "Threats to Submarine Cables"; Wagner, "Submarine Cables and Protections Provided by the Law of the Sea."

38. "The Natives Are Unfriendly (Sometimes)."

39. Vietnam News Briefs, "Vietnam Asks Telcos to Lay More Telecom Lines after Cable Theft."

40. Khan, "Vietnam's Submarine Cable Lost and Found."

41. Carey, "China Says Death Penalty for Damage to Electric Grid."

42. Zahra, "More Internet Cable Incidents Likely to Happen in Mediterranean."

43. Kinyanjui, "Did a Shark Really Bring the Internet Down?"

44. "Of Cables and Conspiracies."

45. Kent Bressie observes that, today, "suddenly, simultaneous cable cuts are assumed to be potential terrorist incidents rather than the results of conflicts with other seabed users" ("Coping Effectively with National-Security Regulation of Undersea Cable").

46. A number of recent texts about the history of specific cable stations exemplify nodal discourse. See, for example, Rowe, *Connecting the Continents*; Tarrant, *Atlantic Sentinel*.

47. Batchen, "Interview with Taryn Simon."

48. For discussion of the visual dimension of infrastructure projects and their ties to colonial projects, see Larkin, *Signal and Noise*, 19.

49. Kilworth, "White Noise," 513.

50. Kilworth, "White Noise," 515.

51. "Via the Microphone," 135.

52. Mattern, "Infrastructural Tourism."

53. Stephenson, "Mother Earth, Mother Board."

54. F. Russell, *A Woman's Journey through the Philippines*. See also Crouch, *On a Surf-Bound Coast* and *Glimpses of Feverland*.

55. Stephenson, "Mother Earth, Mother Board."

56. Stephenson, "Mother Earth, Mother Board."

57. Stephenson, "Mother Earth, Mother Board."

58. Jenkins and Fuller, "Nintendo and New World Narrative."

59. Stephenson, "Mother Earth, Mother Board."

60. Stephenson, "Mother Earth, Mother Board."

61. Stephenson, "Mother Earth, Mother Board."

62. Mattern, "Infrastructural Tourism."

63. Blum, *Tubes*, 279.

64. Blum, *Tubes*, 8.

65. Blum, *Tubes*, 202.

66. Blum, *Tubes*, 279.

THREE. gateway

1. Interview with anonymous cable engineer, October 12, 2010.

2. Taaffe and Gautier, *Geography of Transportation*, 13; Edwards, Jackson, Bowker, and Knobel, "Understanding Infrastructure."

3. Sandvig, "The Internet as Infrastructure."

4. P. Edwards, *The Closed World*.

5. R. Scott, *Gentlemen on Imperial Service*, 25. See also Lobban, "Bamfield Marine Station 1972."

6. R. Scott, "A Short History of the Barkley Sound," 52.

7. H. Taylor, "The First Cable Reached the Island in July 1903."

8. Person, "Life on Midway Island," 259.

9. Hugh Barty-King reports: "Telegraphy, however romantic it may have seemed to

the Press, did not attract young men of any great caliber—and the pay was low, the hours long and, in the case of the remote Porthcurno, tedious. . . . There was a large wastage; many who came were unable to write; some found the techniques beyond them; some had temperaments wildly unsuited; others made themselves unsuitable by over-drinking" (*Girdle round the Earth*, 36).

10. Miller, *Gentlemen of the Cable Service*.

11. See, for example, R. Bruce Scott's description of Pacific Cable Board life in *Gentlemen on Imperial Service* and John Cavalli's cable station newsletter *CaStaNet*.

12. Bernard Finn has documented the contributions of Heart's Content operators to cable transmission technology, including the siphon recorder and duplexing. He argues that this practice had mixed results for operators, as it increased cable transmission speed but also brought more men to work at an already crowded cable station ("Growing Pains at the Crossroads of the World"). The station electrician K. C. Cox, who worked for the Pacific Cable Board, developed the selenium magnifier and the interpolator, two instruments that also helped to improve transmission speed and accuracy ("A New Zealand Paper on Cable Improvements"). A more practical case of innovation included the design of a new diving suit for undersea repair by a Fanning Island cableman ("Cableman Designs a Diving Dress").

13. In places like Broome, Australia, there was so little traffic that cablemen had free time during their shifts.

14. Rowe, *Connecting the Continents*.

15. Rowe, *Connecting the Continents*, 38.

16. Buckland, "Thirty-Four Years' Service in Australia, New Zealand and the East Indies, 1876–1910."

17. Pacific Cable Act 1901, *Account 1907–1908*, 6.

18. H. Taylor, "The First Cable Reached the Island in July 1903."

19. Quoted in Barty-King, *Girdle round the Earth*, 118.

20. Skipper, "Shore=Ends" and "Children of the Submarine Cable Service"; Pacific Cable Board, "Fanning Island-Suva Cables, 1926; Engineer's Final Report, Appendices," August 30, 1927, Museum of Transport and Technology, Auckland, New Zealand.

21. Eastern Telegraph Company, "General Information Relating to Companies' Stations. Rodriguez," Eastern Telegraph Company Records (November 30, 1923): 33, Cable & Wireless Archives, Porthcurno, UK.

22. "Cable Station Dance."

23. Fred Studman, "Recollections of the Good Old Days in Suva in the Mid-Twenties," 210, History of Various Branches and Stations 1966 to 1967, DOC//9/32, Cable & Wireless Archives, Porthcurno, UK.

24. White, "Interview with R. Bruce Scott on December 21, 1973, Victoria B.C."

25. Lobban, "Bamfield Marine Station 1972."

26. Trebett, "Story of 60-Odd Years of the Bamfield Cable Station."

27. "The Latest from Fanning," *Zodiac* 13 (August 1920–July 1921): 365.

28. At other times the Gilbertese would be enticed to put on shows for the cable community. See Adames, "Reminiscences of a Cable Operator."

29. "Norfolk Island, Australia," 115, 116.

30. R. Scott, *Gentlemen on Imperial Service*, 17.

31. Rowe, *Connecting the Continents*, 103.

32. Tarver, "Forty One Years Foreign Service," 107.

33. "Fanning Island," 156–57.

34. One photo in the cable station magazine is titled "A Sample of the Fanning Island Community, Colored and White" and depicts islanders mingling with cablemen as part of the community ("The Latest from Fanning," *Zodiac* 11 [May 1919]: 245).

35. Barty-King, *Girdle round the Earth*, 40–41.

36. Tarver, "Forty One Years Foreign Service," 108.

37. "Cable Thrills," 316.

38. R. Scott, *Gentlemen on Imperial Service*, 17.

39. R. Scott, *Gentlemen on Imperial Service*, 17.

40. Parikka, "Mapping Noise," 267.

41. Miller, *Gentlemen of the Cable Service*, 7.

42. Miller, *Gentlemen of the Cable Service*, 7.

43. "The 'Zodiac' as Geography Primer for Schools."

44. Quoted in "The Gooney Clarion," 67.

45. Quoted in Rowe, *Connecting the Continents*, 60.

46. Skipper, "Children of the Submarine Cable Service," 12.

47. "Roebuck Bay, Broome," 36; "A Volcano in the Philippines," 106.

48. Harwood, "Midway Island, North Pacific," 310.

49. This was not limited to ethnic Others; articles also poked fun at the cluelessness of women at the station. See, for example, "Local Leave with the Wife."

50. Poltock, "Bolshevism in Cocos"; "Theatricals in the Cocos Islands."

51. "Via the Microphone," 135–36.

52. R. Scott, *Gentlemen on Imperial Service*, 106.

53. Maluc, "The Unregenerate Days," 332.

54. Adames, "Reminiscences of a Cable Operator," 95.

55. This was the one major period of technological change in the telegraph system. See Finn, "Introduction."

56. "Another Picture, Representing 1940."

57. "No Chinamen," 288.

58. "No Chinamen," 288.

59. P. Edwards, *The Closed World*, 12.

60. Porthcurno Telegraph Museum, "Nerve Center of Empire."

61. Kelly, "The First Transatlantic Telephone Cable System—Linking Old and New Worlds," 75.

62. "COMPAC—5th Meeting PCMC," p. 24, Record Number 43/7/2, AAMP W3174 909 107, Archives New Zealand, Wellington, New Zealand.

63. "Canadian Overseas Telecommunication Corporation: COMPAC Quarterly Progress Report," New Zealand Works Progress Report to 30/9/61, Archives New Zealand, Wellington, New Zealand.

64. "License Authorizing the Landing and Operation of Two Submarine Cables on the Island of Oahu, Hawaii," FCC 61–258, 99661, File No. S-C-L-20, Trans Pacific Cable

Conference and South East Asia Conference, A1838, 707/49, part 3, National Archives of Australia, Canberra, Australia; "COMPAC—5th Meeting PCMC," 24.

65. "The Terminal Building; COMPAC Cabinet Submissions," AAMF W3174 909 103, 43/7 Archives New Zealand, Wellington, New Zealand.

66. "The Terminal Building; COMPAC Cabinet Submissions," AAMF W3174 909 103, 43/7 Archives New Zealand, Wellington, New Zealand.

67. Pannett and Hercus, "The ANZCAN Submarine Cable System" (unpublished report courtesy of Dave Hercus).

68. "COMPAC—5th Meeting PCMC."

69. Rowe, *Connecting the Continents*, 1.

70. Kelly, "The First Transatlantic Telephone Cable System," 75.

71. Interview with anonymous cable engineer, October 29, 2010.

72. Pacific Cable Management Committee, "Report on Public Relations Meetings, First Series," February 1962, 30, DOC/PCB/1/1 1962, Cable & Wireless Archives, Porthcurno, UK.

73. Cavalli, "Point Arena."

74. "Appendix B, Site Information—Cable Terminal" (September 1959), 1, Pacific Cable Conference 1959, PCC59/16, National Archives of Australia, Canberra, Australia.

75. Interview with anonymous cable consultant, April 4, 2011; interview with anonymous cable engineer, October 29, 2010.

76. "Terrorist Attack Blasts Haifa Office" and "Djakarta Staff Are Safe in Singapore"; Greaves, "Dubai Staff Hijacked in Plane Drama."

77. Many technicians I interviewed are local residents and have worked in the business for at least the past decade, some dating back to the coaxial period.

78. Davies, "Staff," 12.

79. "Training of Engineers from Malaya and Singapore, Paper no. 18, Agenda Item 16," in "COMPAC—5th Meeting PCMC."

80. "Twins at Fanning Island Are Given Silver Cups from the Chairman."

81. "A Career Abroad," 1950, DOC//CW/5/332, Cable & Wireless Archives, Porthcurno, UK.

82. "Singapore Has Trebled Radio and Doubled Cable Capacity."

83. Moss, "SE, JSN, and HK Will Be First Seacom Link"; Moore, "Houng Lee and Waiyamani Win Gold Medals at Pacific Games"; Parkinson, "In England I Was Made to Feel at Home."

84. Lennon, "Fanning Island Copra-Cutters Have Become Experts in Concrete Construction," 13.

85. Quoted in "Hails Cable System."

86. "SEACOM Sta. Is Dedicated."

87. Joyce, "Governor of Fiji Makes History in World of Communications."

88. Quoted in Diotte, "Technologically into the 21st Century."

89. "Teleglobe Raises Building Figures."

90. For example, to commemorate the "birthday" of the Canadian Overseas Telecommunications Corporation in 1975, residents were given tours of the building, in which they were shown the equipment room, power room, and storage depot, as well as

twenty-five exhibits that demonstrated the company's abilities in cable operation and repair ("COTC Marks 25 Years").

91. Interview with anonymous cable station technician, June 22, 2009.

92. Under the consortium model, national telecommunications companies often did not need to put in security except for reliability engineering or to meet government requests.

93. Interview with anonymous cable operations manager, May 26, 2011.

94. Rauscher, *ROGUCCI Study Final Report*, 61.

95. Interview with anonymous cable operations manager, May 26, 2011.

96. Rauscher, *ROGUCCI Study Final Report*, 92.

97. Interview with anonymous cable engineer, May 23, 2011.

98. Interview with anonymous cable enterpreneur, October 1, 2010.

99. Interview with anonymous cable station technician, June 22, 2009.

100. Interview with anonymous cable operations manager, May 26, 2011.

101. Interview with anonymous cable operations manager, May 26, 2011.

102. Interview with anonymous cable engineer, October 12, 2010.

103. Interview with anonymous cable entrepreneur, October 25, 2010.

104. Interview with anonymous cable entrepreneur, October 25, 2010.

105. Rauscher, *ROGUCCI Study Final Report*, 61.

106. Interview with anonymous cable station manager, May 24, 2011.

107. Interview with anonymous cable entrepreneur, October 1, 2010.

108. Interview with anonymous cable station manager, May 24, 2011.

109. Interview with anonymous cable station manager, May 24, 2011.

110. Interview with anonymous cable station manager, May 24, 2011.

111. Rauscher, *ROGUCCI Study Final Report*, 131.

112. Interview with anonymous cable operations manager, May 26, 2011.

113. Interview with anonymous cable operations manager, May 26, 2011.

114. A network operations center is a site where cable workers are concerned with the equipment and system functionality, and a network management center is where cable workers are concerned with routing and restoring traffic. These centers can be located in the same facility.

115. Even if a worker were sitting in the cable station 24/7, these people also might be unreliable. Compared to how long it takes a cable ship to sail to a break, the time it takes to get a person to a station is minimal, especially when there is a network management center that can instantly respond and restore traffic via other networks.

116. Interview with anonymous cable station technician, October 19, 2010.

117. Interview with anonymous cable operations manager, May 26, 2011.

118. Interview with anonymous cable operations manager, May 26, 2011.

119. Interview with anonymous cable operations manager, May 26, 2011.

120. Interview with anonymous cable engineer, October 12, 2010.

121. Interview with anonymous cable operations manager, May 26, 2011.

122. Interview with anonymous cable operations manager, May 26, 2011.

123. Interview with anonymous cable operations manager, May 26, 2011.

124. Interview with anonymous cable engineer, May 23, 2011.

125. Interview with anonymous cable station technician, October 19, 2010.

126. In this way, cable technicians function as boundary workers—who, as Greg Downey observes, "knit disparate communications networks together on a daily basis" ("Virtual Webs, Physical Technologies, and Hidden Workers," 211).

127. Interview with anonymous cable operations manager, May 26, 2011.

128. Interview with anonymous cable operations manager, May 26, 2011.

129. Quoted in H. Barty-King, *Girdle round the Earth.*

130. V. Hughes, "Cable Connections."

131. Beck, "Submarine Cable Workshop."

132. Interview with Dean Veverka, chairman, International Cable Protection Committee, January 16, 2012.

133. Interview with Fiona Beck, CEO, Southern Cross Cable Network, January 16, 2012.

134. Interview with Fiona Beck, January 16, 2012.

135. Interview with Dean Veverka, January 16, 2012.

136. Interview with Dave Willoughby, director of submarine cable systems development, AT&T, January 15, 2012.

137. Interview with Fiona Beck, January 16, 2012.

138. Interview with U.S. Naval Base community liaison officer, July 1, 2009.

139. Pepperell, "Hunt for New Species Begins in Titahi Bay."

140. Lo, "How Fanning Island Got a Toilet and Other Tales of Development."

141. "La Perouse Cable Station," DOC//9/17, Cable & Wireless Archives, Porthcurno, UK.

142. Along the Commonwealth route, stations have been transformed for tourists, a project at times facilitated by cablemen. R. Bruce Scott, an operator in Bamfield, Canada, promoted tourism in the region, ran a resort, and advocated for the creation of a national park.

143. Barty-King, *Girdle round the Earth*, 152.

FOUR. pressure point

1. Rauscher, *ROGUCCI Study Final Report*, 60.

2. Star, "The Ethnography of Infrastructure," 380.

3. These kinds of entanglements can also occur at the cable station, but they are particularly intense at the cable landing point, where companies have far less control over the environment.

4. Organisation for Economic Co-operation and Development, "Building Infrastructure Capacity for Electronic Commerce Leased Line Development and Pricing," 13.

5. D. Press, *Saving Open Space*, 12.

6. North Star Resources, "Southern Cross Submarine Fiber Optic Cable Landing Project."

7. The data that the industry has gathered from plowed areas shows that in shallow waters the seabed is constantly moving, and even in the deeper areas, biological and physical processes cover up the cable, which is designed to stay down for twenty-five

years. Interview with Lionel Carter, marine environmental advisor to the International Cable Protection Committee, October 28, 2010.

8. Modern cable ships with advanced navigation systems are able to precisely control cable routes, limit slack, and bury cable in shallow coastal waters. From 1959 to 2006, none of the 3,460 undersea cable breaks were due to whales. See Wood and Carter, "Whale Entanglements with Submarine Communications Cables"; Heezen, "Whales Entangled in Deep Sea Cables."

9. Adams, "High Fibre Diet."

10. California State Lands Commission, "AT&T Asia America Gateway Fiber Optic Cable Project Final Environmental Impact Report SCH No. 2007111029," 11.

11. California State Lands Commission, "AT&T Asia America Gateway Fiber Optic Cable Project Final Environmental Impact Report SCH No. 2007111029," 16.

12. Interview with anonymous cable engineer, October 12, 2010. See also Southern Cross Cable Network, "About Southern Cross."

13. Interview with anonymous cable operations manager, May 26, 2011.

14. Interview with anonymous cable engineer, May 24, 2011.

15. Interview with anonymous cable entrepreneur, October 25, 2010.

16. Interview with Lionel Carter, October 28, 2010.

17. "Threatened?"

18. Jokiel, Kolinski, Naughton, and Maragos, "Review of Coral Reef Restoration and Mitigation in Hawaii and the U.S. Affiliated Pacific Islands"; Kolinski, "Analysis of Year Long Success of the Transplantation of Corals in Mitigation of a Cable Landing at Tepungan, Piti, Guam."

19. Interview with anonymous cable consultant, April 4, 2011.

20. Wilson, "Landing and Operation of Submarine Cables in the United States," 69.

21. U.S. Senate, *A Bill to Prevent the Unauthorized Landing of Submarine Cables in the United States.*

22. Coggeshall, *An Annotated History of Submarine Cables and Overseas Radiotelegraphs*, 125.

23. California State Lands Commission, "AT&T Asia America Gateway Fiber Optic Cable Project Final Environmental Impact Report SCH No. 2007111029."

24. Interview with anonymous cable engineer, October 12, 2010.

25. Rauscher, *ROGUCCI Study Final Report*, 30–31.

26. New Zealand's Exclusive Economic Zone and Continental Shelf Bill, for example, is intended to manage environmental effects of development in coastal waters and includes submarine cables in its purview; it states that organizations need express permission to engage in placing, altering, extending, or removing a submarine cable on the seabed.

27. Kieve, *The Electric Telegraph*, 104.

28. Gordon, *A Thread across the Ocean*, 201.

29. Engelhard, "One Hundred and Twenty Years of Change in Fishing Power of English North Sea Trawlers."

30. "American Cables."

31. The company suggested that fishermen be excluded from the narrow cable zone, which would be designated a cable "reserve," an exclusion the company claimed would help fishermen since it would effectively establish a fish preserve. Popular discourse in the period characterized fishermen as "cable wreckers" and "hooligans," calling for government intervention to prevent damage "to valuable cable property." See "Cable Wreckers."

32. United Kingdom Inter-Departmental Committee on Injuries to Submarine Cables, *Report*, ix–x.

33. United Kingdom Inter-Departmental Committee on Injuries to Submarine Cables, *Report*, ix–x.

34. "Correspondence re Damage to Cables by Trawlers Volume 1," DOC/ETC/7/4 (1904–1908). Cable & Wireless Archives, Porthcurno, UK.

35. The Eastern Telegraph Company distributed cable charts to fishermen, and the Great Northern Telegraph Company published a special "notice to mariners" about the cable route. "Correspondence re Damage to Cables by Trawlers Volume 1," DOC/ETC/7/4 (1904–1908). Cable & Wireless Archives, Porthcurno, UK.

36. "Correspondence re Damage to Cables by Trawlers Volume 1," DOC/ETC/7/4 (1904–1908). Cable & Wireless Archives, Porthcurno, UK.

37. "Correspondence re Damage to Cables by Trawlers Volume 1," DOC/ETC/7/4 (1904–1908). Cable & Wireless Archives, Porthcurno, UK.

38. "Correspondence re Damage to Cables by Trawlers Volume 1," DOC/ETC/7/4 (1904–1908). Cable & Wireless Archives, Porthcurno, UK.

39. United Kingdom Inter-Departmental Committee on Injuries to Submarine Cables, *Report*.

40. United Kingdom Inter-Departmental Committee on Injuries to Submarine Cables, *Report*, x.

41. United Kingdom Inter-Departmental Committee on Injuries to Submarine Cables, *Report*, x.

42. "Fisheries—Trawlers—Interference with Cables (Submarine). Insertion of Clauses in Licenses Dealing with—Suggests," M1 188, C 357 497, 2/12/282, 1919–1924, Archives New Zealand, Wellington, New Zealand.

43. "Fisheries—Trawlers—Interference with Cables (Submarine)," 1949 part 2.

44. "Fisheries—Trawlers—Interference with Cables (Submarine)," 1949 part 2.

45. "The Eastern Associated Companies' Stand at the Fisheries' Exhibition Causes Unusual Interest amongst the Public," 72.

46. "The Eastern Associated Companies' Stand at the Fisheries' Exhibition Causes Unusual Interest amongst the Public," 73.

47. Letter from W. C. Smith, secretary for marine, to the director-general (overseas telecommunications), General Post Office, July 22, 1949, M.2/12/282, Archives New Zealand, Wellington, New Zealand.

48. Harris, "New Charts for Trawlers Will Cut Fishing Damage," 19.

49. Harris, "New Charts for Trawlers Will Cut Fishing Damage," 22–23.

50. Harris, "New Charts for Trawlers Will Cut Fishing Damage," 22–23.

51. For example, it got the British Admiralty to agree to give information on deep

sea cables to the U.S. Naval Oceanographic Office for oceanographic research vessels' charts ("Admiralty Charts Will Show Cable Routes").

52. Hayes, "Reminiscences of TAT-1."

53. "Exhibitions around the World."

54. These included Australia's Submarine Cables and Pipelines Protection Act of 1963 and New Zealand's Submarine Cables and Pipelines Protection Acts of 1966 and 1977.

55. "'Operation Sea Plow.'"

56. Interview with anonymous cable engineer, May 23, 2011.

57. Harris, "New Charts for Trawlers Will Cut Fishing Damage," 22.

58. Kelly, "The First Transatlantic Telephone Cable System," 82.

59. Interview with anonymous cable engineer, May 24, 2011.

60. One report observed that cables can cost $1–3 million to repair. See New Zealand Parliament, "New Zealand Parliamentary Debate."

61. Rauscher, *ROGUCCI Study Final Report*, 76.

62. In Australia a similar set of arguments were mobilized to update that country's protection act, which was seen as having insufficient incentives for boaters to avoid cables. The act's explanatory memorandum observes: "Any sustained outage would put at risk [AUD] $6.2 billion of GDP derived from international business" (Parliament of the Commonwealth of Australia, "Telecommunications and Other Legislation Amendment"). Recent outages in Australia produced by fishermen—including the disruptions of ANZCAN, Tasman 2, and Southern Cross, and several breaks of SEA-ME-WE 3 off Perth—also helped justify the protection zones.

63. Protection zones were established in Sydney in 2007 around the Southern Cross Cables and the Australia-Japan Cable; the Perth protection zone was established in 2008 around SEA-ME-WE 3. Protection zones were also established in New Zealand around that country's two cable landings at Takapuna and Muriwai. Although fishermen and governments wanted to limit these to a narrow channel, cable companies argued that with all the cables running through a pressure point, it would take much less effort for all communications to be knocked out.

64. Australia Communications and Media Authority, "Submarine Cable Protection Zones."

65. Interview with anonymous cable engineer, May 26, 2011.

66. As another example, Keith Ford-Ramsden and Tara Davenport observe that India has one of the most complicated licensing regimes for undersea cables and has required bonds before foreign systems can land cables, a process that the authors suggest will "have a detrimental effect on route diversity and connectivity of less developed countries" ("The Manufacture and Laying of Submarine Cables," 147).

67. One cable engineer told me that while his company was considering an alternative route for California landing, "We started talking to the fishing industry and they just said, 'No way. You're not gonna run a cable across our fishing grounds. Thank you very much.' And that was the end of it" (interview with anonymous cable engineer, October 12, 2010). The company ended up choosing the existing route instead.

68. Interview with anonymous cable station manager, July 6, 2009; interview with anonymous cable consultant, April 4, 2011.

69. California Coastal Commission, "CDP Application No. E-98–029," 47.

70. Interview with anonymous cable engineer, May 23, 2011.

71. Interview with anonymous cable entrepreneur, October 1, 2010.

72. Interview with anonymous representative of the Naval Seafloor Cable Protection Office, January 16, 2012.

73. His name has been changed.

74. Augé, *Non-Places*.

75. *Mauka* means "toward the mountains" in Hawaiian.

76. *Makai* means "headed toward the sea" in Hawaiian.

77. Morley, "Communications and Transport."

78. Interview with anonymous cable station manager, July 6, 2009.

79. Waianae Neighborhood Board, "Minutes of Regular Meeting, November 6, 2007."

80. Martin, "Tyco's $75M Maili Facility Going for $5M."

81. Martin, "Tyco's $75M Maili Facility Going for $5M"; Waianae Neighborhood Board, "Minutes of Regular Meeting, February 7, 2006."

82. In Bamfield, land that was taken away from indigenous people eventually became cable station property; when the station was done with this land, it was given to the United Church instead of its original owners.

83. White House, NSTAC Chair Letter to President Barack Obama and NSTAC *Response to the Sixty-Day Cyber Study Group* (March 2009): 10–11.

84. Parliament of the Commonwealth of Australia, "Telecommunications and Other Legislation Amendment."

85. Interview with anonymous cable engineer, May 23, 2011.

86. Interview with anonymous cable operations manager, May 26, 2011.

87. Interview with anonymous cable engineer, May 23, 2011.

88. Interview with anonymous cable consultant, April 27, 2011.

FIVE. a network of islands

1. As Ruth Oldenzeil has argued in "Islands," this technopolitical geography helped make islands key nodes of colonial and Cold War power.

2. Lyons, *American Pacificism*, 17.

3. Gillis, *Islands of the Mind*, 2.

4. Royle, *A Geography of Islands*, 11.

5. Hau'ofa, "Our Sea of Islands"; DeLoughery, *Routes and Roots*; Stratford et al., "Envisioning the Archipelago."

6. Galloway and Thacker, *The Exploit*.

7. "First Cable Ended N.Z. Isolation 86 Years Ago," Far North Regional Museum Archives, Kaitaia, New Zealand.

8. Françoise Péron writes: "Today it would seem that nothing any longer distinguishes an island from a section of a contiguous mainland" ("The Contemporary Lure of the Island," 328).

9. Lai, "Discontiguous States of America," 2.

10. Bevacqua, *Chamorros, Ghosts, Non-voting Delegates*, 2 and 80.

11. DeLoughrey, *Routes and Roots*, 8.

12. "Guam Governor Annual Report 1915," Micronesian Area Research Center, Mangilao, Guam.

13. "Guam Governor Annual Report 1915," 18.

14. Cogan, *We Fought the Navy and Won*, 3.

15. Headrick, "Strategic and Military Aspects of Submarine Telegraph Cables 1851–1945," 194.

16. Cogan, *We Fought the Navy and Won*, 16.

17. "Guam Governor Annual Report 1904," 1, Micronesian Area Research Center, Mangilao, Guam.

18. "Guam Governor Annual Report 1904," 1.

19. "Guam Governor Annual Report 1915," 3.

20. "Guam Governor Annual Report 1915," 4.

21. "Guam Governor Annual Report 1917," 31, Micronesian Area Research Center, Mangilao, Guam; Farrell, *The Pictorial History of Guam*, 169.

22. "Guam Governor Annual Report 1955," Micronesian Area Research Center, Mangilao, Guam.

23. Interview with anonymous cable station manager, June 2009.

24. According to the statistics on "Greg's Cable Map."

25. "Late Frank Rostier"; "Levuka's Wireless School."

26. In the 1960s about one in thirteen of Fiji's international transmissions were extended to other Pacific Islands. Fiji Government, *Fiji*, 109.

27. Interview with anonymous cable engineer, May 24, 2011.

28. This was possible in Fiji because telecommunications distribution and technology training were already present. See "Beaming in the Games"; "30 Years on, TV's Unresolved."

29. Fiji International Communications, *Fintel Report and Accounts 1999*, 6.

30. Interview with Ioane Koroivuki, CEO of FINTEL, July 27, 2010.

31. Interview with anonymous cable engineer, October 12, 2010.

32. Fiji International Communications, *Fintel Report and Accounts 2000*, 9.

33. Some countries in similar situations responded by attempting to tax cable traffic, which quickly deterred cable companies from further investment in those countries. One cable manager told me that when Norfolk Island tried to do this, his company's response was: "Well, we can just close down" (interview with anonymous cable manager, May 24, 2011). The island government quickly gave in, as the cable was the country's primary form of external communication.

34. Fiji International Communications, *Fintel Report and Accounts 2000*, 10.

35. Undersea cables' potential to support digital outsourcing and local start-ups has generated interest across the Pacific. Cable builders often believe that if they simply land a new link in a country and provide a high-speed Internet connection, it will entice Pacific islanders (and others) to start new businesses there instead of migrating to Australia, New Zealand, or the United States.

36. Interview with Zarak Khan, manager of industrial zones, Fiji Trade and Investment Bureau, August 4, 2011.

37. Interview with Florence Swamy, acting CEO of the FAVC, August 18, 2010.

38. Fiji has historically been chosen over other islands as a location for production because of its connections with air and sea transportation. *The Blue Lagoon* (Frank Launder, United Kingdom, 1949) was shot in Fiji's Yasawa Islands in part due to their proximity to established transportation routes. It was impossible to ship the exposed negatives from shooting in more remote destinations, such as Tahiti, back to California within fourteen days, the time period within which color film had to be developed. See "Blue Lagoon Party on Way to Fiji"; "TOA Flying Boat Passengers in Fiji."

39. John Connell has examined the importance of the image of isolation for the marketing of FIJI water, one of the country's biggest exports ("'The Taste of Paradise'").

40. Interview with CEO of Tourism Fiji, August 25, 2010.

41. This was made possible by the Century of Aviation Reauthorization Act, which made Micronesia eligible for the U.S. Airport Improvement Program funds.

42. Geddes, "The Mandate for Yap."

43. Geddes, "The Mandate for Yap."

44. Geddes, "The Mandate for Yap," 33.

45. Interview with anonymous cable entrepreneur, October 25, 2010.

46. Interview with Brett O'Riley, October 1, 2010.

47. While Pohnpei benefits from this military extension, if these links are turned off, locals lose access. TE SubCom, "Micronesian Telecommunications Providers."

48. Interview with anonymous cable entrepreneur, October 25, 2010.

49. Interview with Brett O'Riley, October 1, 2010.

50. Interview with Brett O'Riley, October 1, 2010.

51. Interview with anonymous cable engineer, May 24, 2011.

52. Esselaar, Gillwald, and Sutherland, "The Regulation of Undersea Cables and Landing Stations."

53. Interview with anonymous cable manager, October 14, 2010.

54. Big Island Video News.Com, "Undersea Cable Strengthens Hawaii, Tahiti Link"; Pacific Network, "Honotua Pt. 1." The landing was also covered on Hawai'i Public Radio.

55. Interview with Thierry Hars, Honotua's technical project manager in Office des Postes et Télécommunications, November 5, 2011.

56. Hau'ofa, "Our Sea of Islands," 12–13.

57. Interview with Thierry Hars, November 5, 2011.

58. Interview with Thierry Hars, November 5, 2011.

59. Interview with Thierry Hars, November 5, 2011.

60. Interview with Thierry Hars, November 5, 2011.

61. Interview with John Hibbard, April 4, 2011.

SIX. cabled depths

1. Druehl, "Bamfield Survey, Report No. 1."

2. White, "Interview with R. Bruce Scott on December 21, 1973, Victoria B.C."

3. Deacon, *Scientists and the Sea, 1650–1900*, 283.

4. Quoted in Gordon, *A Thread across the Ocean*, 38. See also Field, *The Story of the Atlantic Cable*, 19–20.

5. Rozwadowski, *Fathoming the Ocean*, 14.

6. Deacon warns of simply crediting cable work for the development of marine science, pointing out that many deep-sea soundings were made before cable laying began and that there is little actual evidence that cable laying directly resulted in the scientific study of the oceans (*Scientists and the Sea 1650–1900*, 298).

7. Deacon, *Scientists and the Sea, 1650–1900*, 337.

8. Rozwadowski, *Fathoming the Ocean*, 5–6. Sabine Höhler dates the founding of oceanic sciences to the 1850s ("A Sound Survey").

9. Rozwadowski, *Fathoming the Ocean*, 89–90.

10. After the 1866 transatlantic cable was laid, Rozwadowski observes that the deep sea was not featured nearly as often in popular discussions, having been successfully framed as a habitable environment for undersea cables (Rozwadowski, *Fathoming the Ocean*, 27).

11. Deacon, *Scientists and the Sea, 1650–1900*, 296; Van Aken, "Dutch Oceanographic Research in Indonesia in Colonial Times," 36.

12. "River under Sea Bed."

13. "River under Sea Bed," 335.

14. "The Infinite Variety of Uses for a Cable Ship."

15. "Notes on Marine Animals and Their Preservation," 314.

16. Simpson, "*Electra* Looks for Marine Life."

17. Simpson, "*Electra* Looks for Marine Life," 10.

18. Toyn, "Globigerinella Aquilateralis."

19. Simpson, "*Electra* Looks for Marine Life."

20. Stachurski, *Finding North America*.

21. Hamilton, *Proceedings of the New Zealand Institute 1906*, 52.

22. Hamilton, *Proceedings of the New Zealand Institute 1906*, 60.

23. Pacific Cable Board, "Fanning Island-Suva Cables, 1926."

24. Müeller-Pohl, "Colonialism, Decolonization and the Global Media System."

25. Deacon, *Scientists and the Sea, 1650–1900*, xxxii.

26. Doel, Levin, and Marker, "Extending Modern Cartography to the Ocean Depths."

27. Doel, Levin, and Marker, "Extending Modern Cartography to the Ocean Depths."

28. "The Navigational Aspects of Laying a Submarine Cable in the Tasman Sea," Pacific Cable Conference—Documents, 1838, 707/49 ANNEX, National Archives of Australia, Canberra, Australia; Press and Stringfellow, "The Second Trans-Tasman Cable System," 334.

29. Finn, "Introduction."

30. Turbin, "C.S. Recorder to Carry Out Mediterranean Survey."

31. Burton, "Cable Laying."

32. Burton, "Cable Laying," 17.

33. Burton, "Cable Laying," 17.

34. Hider, "Underwater Camera Developed by Company," 16.

35. Clive Ferguson observes that before the 1950s almost all ROVs were "massive, towed vehicles" used for cable burial ("Subsea Robots," 23).

36. Turbin, "C.S. Recorder to Carry Out Mediterranean Survey," 15.

37. Turbin, "C.S. Recorder to Carry Out Mediterranean Survey," 15.

38. Christ and Wernli, *The ROV Manual.*

39. Ferguson, "Subsea Robots," 27.

40. Jenner, "SCARAB," 5.

41. AT&T, *Submarine Cable Systems Development.*

42. Brooks, "Does This Photograph Show an Old Cable?"

43. "Previewing a New Line of Underwater Television Equipment."

44. Pratt, "Another Country," 12.

45. Interview with Dave Willoughby, January 15, 2012.

46. Interview with Dave Willoughby, January 15, 2012.

47. Interview with Dave Willoughby, January 15, 2012.

48. Interview with anonymous cable entrepreneur, October 25, 2010.

49. Interview with Dave Willoughby, January 15, 2012.

50. Interview with Dave Willoughby, January 15, 2012.

51. Even for its networks today, the U.S. Navy continues to rely on cable technologies, and developments in telecommunications continue to benefit it.

52. Polmar, *The Naval Institute Guide to the Ships and Aircraft of the U.S. Fleet*, 565.

53. Whitman, "SOSUS."

54. Silverstein, "CAESAR, SOSUS, and Submarines."

55. Silverstein, "CAESAR, SOSUS, and Submarines."

56. Flanagan, "A New Way to Use Old Cables."

57. "Scientists Will Visit Fanning."

58. Atoll Institute, "Tabuaran (Fanning) Atoll."

59. Marshall, "Cabies Can Help Scientists Forecast Lethal Tidal Waves."

60. Marshall, "Cables Can Help Scientists Forecast Lethal Tidal Waves," 15.

61. Chave, Duennebier, and Butler, "Putting H2O in the Ocean."

62. The seismological community helped generate some of the early ideas for and discussion about the scientific use of undersea cables. Seismic research continues to be a major motivating force for funding these networks, as governments generally understand the need for better earthquake and tsunami detection.

63. Interview with Verena Tunnicliffe, director of the Victoria Experimental Network under the Sea (VENUS), August 5, 2011.

64. Silverstein, "CAESAR, SOSUS, and Submarines," 409.

65. Interview with Verena Tunnicliffe, August 5, 2011.

66. Chave, Duennebier, and Butler, "Putting H2O in the Ocean."

67. Interview with Verena Tunnicliffe, August 5, 2011.

68. Best, Barnes, Bornhold, and Juniper, "Integrating Continuous Observatory Data."

69. The ALOHA Cabled Observatory was launched in 2011, using the old Hawaii-4 telecommunications cable. The European Sea Floor Observatory Network proposed the

CELTNET for the eastern Atlantic. The Dense Oceanfloor Network System for Earthquakes and Tsunamis (DONET) was established off the coast of Japan, and the Marine Cable Hosted Observatory (MACHO) was set up in Taiwan.

70. Best, Barnes, Bornhold, and Juniper, "Integrating Continuous Observatory Data."

71. Interview with Martin Taylor, president and CEO of Ocean Networks Canada, August 4, 2011.

72. Helmreich, *Alien Ocean*, 243.

73. Interview with Verena Tunnicliffe, August 5, 2011.

74. Interview with Verena Tunnicliffe, August 5, 2011.

75. Interview with Verena Tunnicliffe, August 5, 2011.

76. Interview with Mairi Best, associate director for science, NEPTUNE Canada, August 24, 2011.

77. You and Howe, "Turning Submarine Telecommunications Cables into a Real-Time Multi-Purpose Global Climate Change Monitoring Network," 185.

78. Golovchenko, "Dual Purpose Subsea Cables."

79. Golovchenko, "Dual Purpose Subsea Cables."

80. Golovchenko, "Dual Purpose Subsea Cables."

81. Bressie, "Gathering Environmental Data Using Submarine Cables."

82. "Cable Communications and Control Systems for Offshore Applications."

83. The first oil industry ROV was the Shell Mobot, used in 1962 to assist in drilling activities off the California coast. In the mid-1970s, special-purpose ROVs were used to mine manganese in the ocean. See Ferguson, "Subsea Robots."

84. Mariano and Baroni, "Enabling Fiber Optic Undersea Cable Technologies for Enhanced Offshore Petroleum Industry Telecommunication Applications," 280.

85. Munier and Haaland, "BP GoM."

86. Mariano and Baroni, "Enabling Fiber Optic Undersea Cable Technologies," 280.

87. Mariano and Baroni, "Enabling Fiber Optic Undersea Cable Technologies."

88. Interview with anonymous representative of the Naval Seafloor Cable Protection Office, January 16, 2012.

89. Griesemer, *Signal & Noise*, 32.

90. The shore end is often the more important part to remove, since cables are funneled through a relatively narrow space; they are often removed to prevent conflict with subsequent systems.

91. Interview with anonymous cable engineer, May 24, 2011.

92. Interview with anonymous representative of the Ocean City Reef Foundation, April 30, 2010.

93. Department of Agriculture, Fisheries, and Forestry of the Isle of Man, "Port Erin 'Cable Reef' Proposal."

94. Vietnam News Briefs, "Vietnam Asks Telcos to Lay More Telecom Lines after Cable Theft."

conclusion

1. Quoted in Shimura, *International Submarine Cable Systems*, 8.

2. Parks, "Around the Antennae Tree." For a detailed discussion of modes of infrastructural intelligibility, see Parks, "Technostruggles and the Satellite Dish."

3. Carter acknowledges that education is a certainly a "double-edged sword" that could reveal sensitive information to potential cable enemies, and these potential consequences always need to be considered. Interview with Lionel Carter, marine environmental advisor, International Cable Protection Committee, October 28, 2010.

4. Currently, less than 5 percent of the billions of dollars spent to fund cable systems comes from governments, yet their support through policy, regulation, and finance remains critical to establishing robust digital networks. See Burnett, Davenport, and Beckman, "Why Submarine Cables?"

5. Buckley, "Conference Call to the Moon."

6. "The Demands of Space Exploration have Taken STC to the Bottom of the Ocean."

7. "ICECAN."

8. "Undersea Seastation."

9. Bauman, *Liquid Modernity*, 14.

10. De Cauter, *The Capsular Civilization*, 109–10.

BIBLIOGRAPHY

Adames, W. G. "Reminiscences of a Cable Operator." *Zodiac* 29 (August 1937–July 1938): 90–95.

Adams, Ruth. "High Fibre Diet." *Zodiac* 35 (1987): 8–11.

"Admiralty Charts Will Show Cable Routes." *Zodiac* 56 (May 1964): 11.

Ahvenainen, Jorma. *The Far Eastern Telegraphs: The History of Telegraphic Communications between the Far East, Europe, and America before the First World War.* Helsinki: Suomalainen Tiedeakatemia, 1981.

"American Cables: Alleged Damage by British Trawlers." *Daily Telegraph*, May 21, 1908.

"Another Picture, Representing 1940." *Zodiac* 22 (August 1929–July 1930): 61.

AT&T. *Submarine Cable Systems Development.* United States: AT&T, 1958.

Atoll Institute. "Tabuaran (Fanning) Atoll." Accessed July 6, 2014. http://archive.today/IWva3.

Augé, Marc. *Non-Places: Introduction to an Anthropology of Supermodernity.* London: Verso, 2009.

Australia Communications and Media Authority. "Submarine Cable Protection Zones." Accessed July 7, 2014. http://www.acma.gov.au/Industry/Telco/Infrastructure/Submarine-cabling-and-protection-zones/submarine-telecommunications-cables-submarine-cable-zones-i-acma.

B., H. F. "Cablemen." *Zodiac* 9 (June 1916–June 1917): 211.

Baran, Paul. "On Distributed Communications." Santa Monica, CA: RAND, 1964.

Barty-King, Hugh. *Girdle round the Earth: The Story of Cable and Wireless and Its Predecessors to Mark the Group's Jubilee 1929–1979.* London: Heinemann, 1979.

Batchen, Geoffrey. "Interview with Taryn Simon." *Museo* 8 (2008). Accessed July 6, 2014. http://www.museomagazine.com/TARYN-SIMON.

Bauman, Zygmunt. *Liquid Modernity.* Cambridge: Polity, 2000.

"Beaming in the Games." *Islands Business*, September 1984, 59.

Beaufils, J. M., and J. L. Chabert. "Interconnection of Future Submarine Cables in the Global Communications Web." In *Proceedings of the 1997 SubOptic Conference*, 91–97. San Francisco: SubOptic, 1997.

Beck, Fiona. "Submarine Cable Workshop." Presentation at the 2010 Pacific Telecommunications Conference, Honolulu, HI, January 17.

Best, Mairi, Christopher Barnes, Brian Bornhold, and Kim Juniper. "Integrating Continuous Observatory Data: A Multidisciplinary View of the Ocean in Four Dimensions." In *Seafloor Observatories: A New Vision of the Earth from the Abyss*, edited by Paolo Favali, Angela de Santis, and Laura Beranzoli. Berlin: Springer, 2014.

Bevacqua, Michael Lujan. *Chamorros, Ghosts, Non-voting Delegates: GUAM! Where the Production of America's Sovereignty Begins*. Ph.D dissertation, Ethnic Studies, University of California San Diego, 2010.

Big Island Video News.Com. "Undersea Cable Strengthens Hawaii, Tahiti Link." March 2, 2010. Accessed July 5, 2012. http://www.bigislandvideonews.com/2010/03/02/video-undersea-cable-strengthens-hawaii-tahiti-link/.

Blanchette, Jean-François. "A Material History of Bits." *Journal of the American Society for Information Science and Technology* 62, no. 6 (2011): 1042–57.

"Blue Lagoon Party on Way to Fiji." *Pacific Islands Monthly*, December 1947, 6.

Blum, Andrew. *Tubes: A Journey to the Center of the Internet*. New York: HarperCollins, 2012.

Bozak, Nadia. *The Cinematic Footprint: Lights, Camera, Natural Resources*. New Brunswick, NJ: Rutgers University Press, 2012.

"Breaking News." *Submarine Telecoms Forum* 1, no. 4 (2001): 5.

Bressie, Kent. "Coping Effectively with National-Security Regulation of Undersea Cable." Presentation at SubOptic 2013, Paris, April 24, 2013.

———. "Gathering Environmental Data Using Submarine Cables: The Non-Technical Issues." Presentation at the Pacific Telecommunications Council Conference, Honolulu, HI, January 17, 2012.

Bressie, Kent, and Madeleine Findley. "Cybersecurity Developments Raise Growing Regulatory Concerns for Undersea Cable Industry." *Submarine Telecoms Forum* 69 (March 2013): 8–16.

Brewin, Bob. "Northwest Passage Fiber Optic Line Could Support Defense Arctic Strategy." *Nextgov.com*. November 27, 2013, Accessed July 7, 2014. http://www.nextgov.com/defense/2013/11/northwest-passage-fiber-optic-line-could-support-defense-arctic-strategy/74639.

Brooks, N. J. "Does This Photograph Show an Old Cable?" *Zodiac* 56 (January 1964).

Buckland, R. G. S. "Thirty-Four Years' Service in Australia, New Zealand and the East Indies, 1876–1910." *Zodiac* 17 (January 1925–December 1925): 230–34.

Buckley, Edmond C. "Conference Call to the Moon." *Zodiac* 58 (October 1966): 8–9.

Burnett, Douglas R., Tara M. Davenport, and Robert C. Beckman. "Why Submarine Cables?" In *Submarine Cables: The Handbook of Law and Policy*, edited by Douglas R. Burnett, Robert C. Beckman, and Tara M. Davenport, 1–15. Leiden, the Netherlands: Martinus Nijhoff, 2014.

Burns, Bill. "History of the Atlantic Cable and Undersea Communications From the First Submarine Cable of 1850 to the Worldwide Fiber Optic Network." *AtlanticCable.com*.

Burton, Robin. "Cable Laying: There's a Fine Art in Throwing a Fortune Overboard." *Zodiac* 22 (1979): 16–19.

"Cable Communications and Control Systems for Offshore Applications." *Underseas Cable World* 2, no. 4 (1969).
"The Cable Companies and the War." *Zodiac* 8 (June 1914–June 1915): 236–37.
"Cable Station Dance." *Guam News Letter*, August 1917.
"Cable Thrills." *Zodiac* 17 (January–July 1926): 316.
"Cable Wreckers: Hooligan Fishermen Who Hold Up Commerce." *Daily Express*, May 26, 1908.
"Cableman Designs a Diving Dress." *Sydney Morning Herald*, September 16, 1911.
"Cables in the Pacific." *Underseas Cable World* 2, no. 5 (1969).
California Coastal Commission. "CDP Application No. E-98-029." San Francisco: California Coastal Commission, 2000.
California State Lands Commission. "AT&T Asia America Gateway Fiber Optic Cable Project Final Environmental Impact Report SCH No. 2007111029." San Francisco: California State Lands Commission, 2009. Accessed July 14, 2014. http://www.slc.ca.gov/division_pages/DEPM/Reports/ATT/ATT.html.
"*Call the World* to Be Shown at Two Major Film Festivals." *Zodiac* (June 1962): 18.
Carey, John. "China Says Death Penalty for Damage to Electric Grid." *OpEdNews.com*. July 14, 2015. http://www.opednews.com/articles/opedne_john_car_070821_china_says_death_pen.htm.
Carter, Samuel. *Cyrus Field: A Man of Two Worlds*. New York: G. P. Putnam's Sons, 1968.
Castells, Manuel. "A Network Theory of Power." *International Journal of Communication* 5 (2011): 773–87.
Cavalli, John. "Point Arena: From Lighthouse to Lightguide." 1995. Accessed April 2, 2010. http://www.privateline.com/cablestation/index.html.
Chave, Alan D., Fred K. Duennebier, and Rhett Butler. "Putting H2O in the Ocean: The Hawaii-2 Observatory Is the First Long-Term, Mid-Ocean Seafloor Observatory." *Oceanus* 42, no. 1 (2000): 6–9.
Choi, Junho H., George A. Barnett, and Bum-So Chon. "Comparing World-City Networks: A Network Analysis of Internet Backbone and Air Transport Inter-City Linkages." *Global Networks* 6, no. 1 (2006): 81–99.
Christ, Robert D., and Robert L. Wernli. *The ROV Manual: A User Guide to Observation-Class Remotely Operated Vehicles*. Amsterdam: Butterworth-Heinemann, 2007.
Chun, Wendy. *Control and Freedom: Power and Paranoia in the Age of Fiber Optics*. Cambridge, MA: MIT Press, 2006.
Clarke, Arthur C. *Voice across the Sea*. New York: Harper and Brothers, 1958.
Cogan, Doloris Coulter. *We Fought the Navy and Won: Guam's Quest for Democracy*. Honolulu: University of Hawaiʻi Press, 2008.
Coggeshall, Ivan S. *An Annotated History of Submarine Cables and Overseas Radiotelegraphs: 1851–1934 with Special Reference to the Western Union Telegraph Company*. Edited by Donard de Cogan. University of East Anglia School of Info.Systems, 1993.
Collins, Richard. "The Bermuda Agreement 1945." CRESC Working Paper Series, Working Paper No. 99. Buckinghamshire, UK: Centre for Research on Socio-Cultural Change, the Open University, May 2010.

"Commentary on a Premature Burial (at Sea)." *Underseas Cable World* 1, no. 8 (1968).
Connell, John. "'The Taste of Paradise': Selling Fiji and FIJI Water." *Asia Pacific Viewpoint* 47, no. 3 (2006): 342–50.
Cookson, Gillian. *The Cable: The Wire That Changed the World*. Stroud, UK: Tempus, 2003.
Costin, Michael. "The Impact of Australian Telecommunications Deregulation upon International Cable Planning." In *Proceedings of the 1997 SubOptic Conference*, 59–65. San Francisco: SubOptic, 1997.
"COTC Marks 25 Years." *Alberni Valley Times*, December 15, 1975.
Crouch, Arthur P. *Glimpses of Feverland: or, A Cruise in West African Waters*. London: Sampson Low, 1889.
———. *On a Surf-Bound Coast: or, Cable-Laying in the African Tropics*. London: Sampson Low, 1887.
Cubbitt, Sean, Hassan, Robert, and Volkmer, Ingrid. "Does Cloud Computing Have a Silver Lining?" *Media, Culture & Society* 33, no.1 (2011): 149–58.
Cunningham, Douglas. "Can the Arctic Provide an Alternative Route?" Paper presented at the 2012 Pacific Telecommunications Conference, Honolulu, HI, January 17.
Curtin, Michael. "Media Capitals: Cultural Geographies of Global TV." In *Television after TV: Essays on a Medium in Transition*, edited by Lynn Spigel and Jan Olsson, 270–302. Durham, NC: Duke University Press, 2004.
Dadouris, L. S., M. C. Singhi, and S. H. Long. "The Impact of the Internet and Broadband Service Offerings on Submarine System Cable Capacity into the 21st Century." In *Proceedings of the 1997 SubOptic Conference*, 48–54. San Francisco: SubOptic, 1997.
Davies, W. H. "Staff." *Zodiac* 59 (February 1967): 12–13.
Deacon, Margaret. *Scientists and the Sea, 1650–1900: A Study of Marine Science*. 2nd ed. Aldershot, UK: Ashgate, 1997.
De Cauter, Lieven. *The Capsular Civilization: On the City in the Age of Fear*. Rotterdam, the Netherlands: NAi, 2004.
De Certeau, Michel. *The Practice of Everyday Life*. Translated by Stephen F. Rendall. Berkeley: University of California Press, 1984.
De Cogan, Donard. "Background to the 1858 Telegraph Cable." Paper presented at the Institution of Engineering and Technology Seminar on the Story of Transatlantic Communications, Manchester, UK, October 28, 2008.
Dellinger, Matt. "The Million Dollar Millisecond." *Popular Science Magazine*, May 2011, 33.
DeLoughrey, Elizabeth. *Routes and Roots: Navigating Caribbean and Pacific Island Literatures*. Honolulu: University of Hawai'i Press, 2007.
"The Demands of Space Exploration Have Taken STC to the Bottom of the Ocean." *Zodiac* 59 (August 1967).
Department of Agriculture, Fisheries, and Forestry of the Isle of Man. "Port Erin 'Cable Reef' Proposal: Summary of Responses to the Consultation." Douglas: Department of Agriculture, Fisheries, and Forestry of the Isle of Man, July 30, 2009.
Devos, Jean. "Changing Relationships and Responsibilities: Towards the Networks of

the 21st Century." In *Proceedings of the 1997 SubOptic Conference*, 211. San Francisco: SubOptic, 1997.
Dibner, Bern. *The Atlantic Cable*. Norwalk, CT: Burdy Library, 1959.
Diotte, Rob. "Technologically into the 21st Century." *Alberni Valley Times*, October 18, 1983.
"Djakarta Staff Are Safe in Singapore." *Zodiac* 55 (July 1963): 10.
Djelic, Marie-Laure, and Sigrid Quack. "Overcoming Path Dependency: Path Generation in Open Systems." *Theory and Society* 36, no. 2 (2007): 161–86.
Dodge, Martin, and Rob Kitchin. *Mapping Cyberspace*. London: Routledge, 2000.
Doel, Ronald E., Tanya J. Levin, and Mason K. Marker. "Extending Modern Cartography to the Ocean Depths: Military Patronage, Cold War Priorities, and the Heezen-Tharp Mapping Project, 1952–1959." *Journal of Historical Geography* 32, no. 3 (2006): 605–26.
Doesburg, Anthony. "Across the Tasman, Two Cables Better Than One." *New Zealand Herald*, April 25, 2008.
Dourish, Paul, and Genevieve Bell. *Divining a Digital Future: Mess and Mythology in Ubiquitous Computing*. Cambridge, MA: MIT Press, 2011.
———. "The Infrastructure of Experience and the Experience of Infrastructure: Meaning and Structure in Everyday Encounters with Space." *Environment and Planning B: Planning and Design* 34, no. 3 (2007): 414–30.
Downey, Greg. "Virtual Webs, Physical Technologies, and Hidden Workers: The Spaces of Labor in Information Internetworks." *Technology and Culture* 42, no. 2 (2001): 209–35.
Downey, Patrick. *Pacific Wiretap*. Bloomington, IN: iUniverse.com, 2008.
Druehl, Louis D. "Bamfield Survey, Report No. 1." *Bamfield Marine Station*, April 1970.
"East of Vancouver." *Zodiac* 31 (August 1939–July 1940): 126–30.
"The Eastern Associated Companies' Stand at the Fisheries' Exhibition Causes Unusual Interest amongst the Public." *Zodiac* 15 (September 1922): 72–73.
Edwards, John H. "Deep Sea Interlude." *Zodiac* 29 (August 1937–July 1938): 367.
Edwards, Paul N. *The Closed World: Computers and the Politics of Discourse in Cold War America*. Cambridge, MA: MIT Press, 1997.
Edwards, Paul N., Steven J. Jackson, Geoffrey C. Bowker, and Cory P. Knobel. "Understanding Infrastructure: Dynamics, Tensions, and Design." Report of a Workshop on "History & Theory of Infrastructure: Lessons for New Scientific Cyberinfrastructures." January 2007. Accessed July 14, 2014. http://deepblue.lib.umich.edu/bitstream/handle/2027.42/49353/UnderstandingInfrastructure2007.pdf.
Elsaesser, Thomas. "'Constructive Instability', or: The Life of Things as the Cinema's Afterlife?" In *Video Vortex Reader: Responses to YouTube*, edited by Geert Lovink and Sabine Niederer, 13–32. Amsterdam: Institute of Network Cultures, 2008.
Engelhard, Georg H. "One Hundred and Twenty Years of Change in Fishing Power of English North Sea Trawlers." In *Advances in Fisheries Science 50 Years On from Beverton and Holt*, edited by A. Payne, J. Cotter, and T. Potter. Oxford: Blackwell, 2009.
"Entrepreneurs Announce $900m Fast Broadband Plan." *New Zealand Herald*, March 11, 2010.

Esselaar, Steve, Alison Gillwald, and Ewan Sutherland. "The Regulation of Undersea Cables and Landing Stations." April 24, 2007. Learning Information Networking and Knowledge Centre. Accessed July 13, 2014. http://ssrn.com/abstract=1472405.

"Exhibitions around the World." *Mercury* (1975).

"Fanning Island." *Zodiac* 28 (August 1935–July 1936): 156–57.

Farrell, Don A. *The Pictorial History of Guam: The Sacrifice 1919–1943*. Tamuning, Guam: Micronesian Productions, 1991.

Ferguson, Clive. "Subsea Robots." *Engineering World*, October 1991, 23–29.

Field, Henry M. *The Story of the Atlantic Cable*. New York: Charles Scribner's Sons, 1892.

Fiji Government. *Fiji: Report for the Year, 1966*. Suva: Government Printer, 1967.

Fiji International Communications. *Fintel Report and Accounts 1999*. Suva: Fiji International Communications, 1999.

———. *Fintel Report and Accounts 2000*. Suva: Fiji International Communications, 2000.

Findley, Madeleine V. "Now You See It: Strategies for Ensuring That Regulatory Agencies See and Consult the Undersea Cable Industry as a Key Stakeholder." Paper presented at the International Cable Protection Committee Plenary, Miami, FL, May 22, 2013.

Finn, Bernard. "Growing Pains at the Crossroads of the World: A Submarine Cable Station in the 1870s." *Proceedings of the IEEE* 64, no. 9 (1976): 1287–92.

———. "Introduction." In *Development of Submarine Cable Communications*, edited by Bernard Finn. New York: Arno, 1980.

———. "Submarine Telegraphy: A Study in Technical Stagnation." In *Communications under the Seas: The Evolving Cable Network and Its Implications*, edited by Bernard Finn and Daqing Yang, 9–24. Cambridge, MA: MIT Press, 2009.

Flanagan, Gordon J. "A New Way to Use Old Cables." *Zodiac* 56 (November 1964): 19–22.

Fletcher, Hamish. "Kordia Puts Brakes on Cable." *New Zealand Herald*, September 2, 2011.

———. "New Pacific Cable Link Plan Unveiled." *New Zealand Herald*, September 6, 2012.

Ford-Ramsden, Keith, and Douglas R. Burnett. "Submarine Cable Repair and Maintenance." In *Submarine Cables: The Handbook of Law and Policy*, edited by Douglas R. Burnett, Robert C. Beckman, and Tara M. Davenport, 155–78. Leiden, the Netherlands: Martinus Nijhoff, 2014.

Ford-Ramsden, Keith, and Tara Davenport. "The Manufacture and Laying of Submarine Cables." In *Submarine Cables: The Handbook of Law and Policy*, edited by Douglas R. Burnett, Robert C. Beckman, and Tara M. Davenport, 123–54. Leiden, the Netherlands: Martinus Nijhoff, 2014.

Fuller, Matthew. *Media Ecologies: Materialist Energies in Art and Technoculture*. Cambridge, MA: MIT Press, 2005.

Galloway, Alexander R. "Networks." In *Critical Terms for Media Studies*, edited by

W. J. T. Mitchell and Mark B. N. Hansen, 280–96. Chicago: University of Chicago Press, 2010.
Galloway, Alexander R., and Eugene Thacker. *The Exploit: A Theory of Networks*. Minneapolis: University of Minnesota Press, 2007.
Geddes, David. "The Mandate for Yap." *History Today* 43, no. 12 (1993): 32–37.
Gillis, John. *Islands of the Mind: How the Human Imagination Created the Atlantic World*. New York: Palgrave Macmillan, 2004.
Golovchenko, Ekaterina. "Dual Purpose Subsea Cables—Gathering Scientific Data from the Ocean Floor." Paper presented at the Pacific Telecommunications Council Conference, Honolulu, HI, January 17, 2012.
Gonze, Josh. "U.S./Japan Trade War Broils." *Network World* 4, no. 14 (1987): 6.
"The Gooney Clarion." *Zodiac* 6 (June 1912–May 1913): 67.
Gordon, John Steele. *A Thread across the Ocean: The Heroic Story of The Transatlantic Cable*. New York: Walker, 2002.
Graham, Stephen, ed. *The Cybercities Reader*. New York: Routledge, 2004.
Graham, Stephen, and Simon Marvin. *Splintering Urbanism: Networked Infrastructures, Technological Mobilities and the Urban Condition*. London: Routledge, 2001.
Graham, Stephen, and Nigel Thrift. "Out of Order: Understanding Repair and Maintenance." *Theory, Culture & Society* 24, no. 3 (2007): 1–25.
Greaves, Derrick. "Dubai Staff Hijacked in Plane Drama." *Mercury*, no. 10 (September 1973).
"Greg's Cable Map." Accessed November 24, 2012. www.cablemap.info.
Griesemer, John. *Signal & Noise*. New York: Picador, 2003.
"Hails Cable System." *Pacific Journal*, March 30, 1967.
Hamilton, A. *Proceedings of the New Zealand Institute 1906*. Vol. 39. Wellington: John Mackay, Government Printing Office, 1907.
Harris, Arthur. "New Charts for Trawlers Will Cut Fishing Damage." *Zodiac* (May 1961): 19–23.
Harwood, P. J. "Midway Island, North Pacific." *Zodiac* 5 (May 1912): 308–11.
Hau'ofa, Epeli. "Our Sea of Islands." In *A New Oceania: Rediscovering Our Sea of Islands*, edited by Eric Waddell, Vijay Naidu, and Epeli Hau'ofa, 2–16. Suva, Fiji: University of the South Pacific School of Social and Economic Development, 1993.
Hayes, Jeremiah F. "Reminiscences of TAT-1." IEEE History Center, 2. Accessed July 13, 2014. http://www.ieeeghn.org/wiki/images/8/8e/Hayes.pdf.
Headrick, Daniel R. "Gutta-Percha: A Case of Resource Depletion and International Rivalry." IEEE *Technology and Society Magazine* 6, no. 4 (1987): 12–18.
———. "Strategic and Military Aspects of Submarine Telegraph Cables 1851–1945." In *Communications under the Seas: The Evolving Cable Network and Its Implications*, edited by Bernard Finn and Daqing Yang, 185–208. Cambridge, MA: MIT Press, 2009.
———. *The Tentacles of Progress: Technology Transfer in the Age of Imperialism, 1850–1940*. New York: Oxford University Press, 1988.
———. *The Tools of Empire: Technology and European Imperialism in the Nineteenth Century*. New York: Oxford University Press, 1981.

Hedges, Laura. "Hibernia Networks Reveals New Partner for Project Express Cable Deployment." *Capacity*, May 13, 2013. Accessed July 13, 2014. http://www.capacitymagazine.com/Article/3204973/Hibernia-Networks-reveals-new-partner-for-Project-express-cable-deployment.html.

Heezen, Bruce C. "Whales Entangled in Deep Sea Cables." *Deep Sea Research* 4 (1953): 105–15.

Helmreich, Stefan. *Alien Ocean: Anthopological Voyages in Microbial Seas*. Berkeley: University of California Press, 2011.

Hempstead, Colin A. "Representations of Transatlantic Telegraphy." *Engineering Science and Education Journal* 4, no. 6 (December 1995): 17–25.

Hendery, Simon."Strength in Telco Networking Numbers." *New Zealand Herald*, August 11, 2010.

Hider, D. J. "Underwater Camera Developed by Company." *Zodiac* 59 (February 1967): 16–17.

Hill, Edwin C., commentator. "An Underwater Cable and Other Communication News and Views." Accessed August 9, 2009. http://www.youtube.com/watch?v=DM-pLPy8Mdo.

Hills, Jill. *The Struggle for Control of Global Communication: The Formative Century*. Urbana: University of Illinois Press, 2002.

———. *Telecommunications and Empire*. Urbana: University of Illinois Press, 2007.

Höhler, Sabine. "A Sound Survey: The Technological Perception of Ocean Depth, 1850–1930." In *Transforming Spaces: The Topological Turn in Technology Studies*, edited by Mikael Hard, Andreas Losch, and Dirk Verdicchio. Publication of the International Conference held in Darmstadt, Germany, March 22–24, 2002. Accessed July 13, 2014. http://www.ifs.tu-darmstadt.de/fileadmin/gradkoll//Publikationen/space-folder/pdf/Hoehler.pdf.

Holroyd-Doveton, C. Untitled article. *Zodiac* 10 (April 1918): 1.

Hughes, Thomas Parke. *Networks of Power: Electrification in Western Society, 1880–1930*. Baltimore, MD: Johns Hopkins University Press, 1993.

Hughes, Val. "Cable Connections." *Alberni Valley Times*, October 11, 2002.

"ICECAN." *Underseas Cable World* 1, no. 4 (1967).

"The Infinite Variety of Uses for a Cable Ship." *Zodiac* 17 (July 1925): 362.

Jackson, Steven J., Paul N. Edwards, Geoffrey C. Bowker, and Cory P. Nobel. "Understanding Infrastructure: History, Heuristics, and Cyberinfrastructure Policy." *First Monday* 12, no. 6 (June 2007). Accessed July 13, 2014. http://firstmonday.org/ojs/index.php/fm/article/view/1904/1786.

Jenkins, Henry, and Mary Fuller. "Nintendo and New World Narrative." In *Cybersociety: Computer-mediated Communication and Community*, edited by Steve Jones, 57–72. Thousand Oaks, CA: SAGE, 1994.

Jenner, Paul. "SCARAB: The Eye in the Sea." *Zodiac* 27 (1981): 2–5.

Johnson, George. *The All Red Line: The Annals and Aims of the Pacific Cable Project*. Ottawa: James Hope and Sons, 1903.

Jokiel, Paul L., Steven P. Kolinski, John Naughton, and James E. Maragos. "Review of Coral Reef Restoration and Mitigation in Hawaii and the U.S. Affiliated Pacific

Islands." In *Coral Reef Restoration Handbook*, edited by William F. Precht, 271–90. Boca Raton, FL: CRC, 2006.

Joyce, George. "Governor of Fiji Makes History in World of Communications." *Zodiac* 55 (January 1963): 6–7.

Kedar, Michael, and Laurent Duplantie. "Building the International Infrastructure for the Second Tier Market." In *Proceedings of the 1997 SubOptic Conference*, 55–58. San Francisco: SubOptic, 1997.

Kelly, Philip. "The First Transatlantic Telephone Cable System—Linking Old and New Worlds." Paper presented at the Institution of Engineering and Technology Seminar on the Story of Transatlantic Communications, 73–83. Manchester, UK, October 28, 2008.

Kelty, Chris. "Against Networks." 2007. Accessed May 18, 2014. http://kelty.org/or/papers/unpublishable/Kelty.AgainstNetworks.2007.pdf.

Kennedy, Paul M. "Imperial Cable Communications and Strategy, 1870–1914." *English Historical Review* 86, no. 341 (1971): 728–52.

Khan, Abu Saeed. "Vietnam's Submarine Cable Lost and Found." *bdnews24.com*, June 2, 2007. Accessed July 13, 2014. http://bdnews24.com/bangladesh/2007/06/01/vietnam-s-submarine-cable-lost-and-found.

Kieve, Jeffrey L. *The Electric Telegraph: A Social and Economic History*. Newton Abbot, UK: David and Charles, 1973.

Kilworth, Gary. "White Noise." In *Year's Best Fantasy and Horror*, edited by Ellen Datlow, 3: 508–16. New York: St Martin's, 1990.

Kinyanjui, Kui. "Did a Shark Really Bring the Internet Down?" *Business Daily*, February 12, 2008.

Kipling, Rudyard. "The Deep-Sea Cables." In Rudyard Kipling, *The Collected Poems of Rudyard Kipling*, edited by R. T. Jones, 181. Ware, UK: Wordsworth, 1994.

Kirschenbaum, Matthew G. *Mechanisms: New Media and the Forensic Imagination*. Cambridge, MA: MIT Press, 2008.

Knuesel, Ariane. "British Diplomacy and the Telegraph in Nineteenth-Century China." *Diplomacy & Statecraft* 18, no. 3 (2007): 517–37.

Kolinski, Steven P. "Analysis of Year Long Success of the Transplantation of Corals in Mitigation of a Cable Landing at Tepungan, Piti, Guam: 2001–2002." *NOAA Fisheries Service Pacific Islands Regional Office Report*. Honolulu, HI: National Oceanic and Atmospheric Administration National Marine Fisheries Service, 2002.

Lai, Paul. "Discontiguous States of America: The Paradox of Unincorporation in Craig Santos Perez's Poetics of Chamorro Guam." *Journal of Transnational American Studies* 3, no. 2 (2011): 1–28.

Larkin, Brian. *Signal and Noise: Media, Infrastructure, and Urban Culture in Nigeria*. Durham, NC: Duke University Press, 2008.

"Late Frank Rostier." *Pacific Islands Monthly*, April 1941, 48.

"The Latest from Fanning." *Zodiac* 11 (May 1919): 242–45.

"The Latest from Fanning." *Zodiac* 13 (August 1920–July 1921): 364–69.

Lennon, J. B. "Fanning Island Copra-Cutters Have Become Experts in Concrete Construction." *Zodiac* 52 (January 1961): 12–13.

"Levuka's Wireless School." *Pacific Islands Monthly*, March 1946, 38.

Liu, Alan. "Remembering Networks: Agrippa, RoSE, and Network Archaeology." Paper presented at the Network Archaeology Conference, Miami University, Oxford, OH, April 21, 2012.

Liu, Alan, and Scott Pound. "The Amoderns: Reengaging the Humanities: A Feature Interview with Alan Liu." *amodern* 2 (2013).

Lo, Catherine. "How Fanning Island Got a Toilet and Other Tales of Development." *Honolulu Weekly*, November 24, 2004.

Lobban, Chris. "Bamfield Marine Station 1972: A Very Good Year for Small-Town Marine Biologists." *Bamfield Marine Station*, February 1973.

"Local Leave with the Wife." *Zodiac* 25 (August 1932–July 1933): 175.

Lovink, Geert. *Dark Fiber: Tracking Critical Internet Culture*. Cambridge, MA: MIT Press, 2002.

Lyons, Paul. *American Pacificism: Oceania in the U.S. Imagination*. New York: Routledge, 2006.

Mackenzie, Adrian. *Wirelessness: Radical Empiricism in Network Cultures*. Cambridge, MA: MIT Press, 2010.

Main, Linda. "The Global Information Infrastructure: Empowerment or Imperialism?" *Third World Quarterly* 22, no. 1 (2001): 83–97.

Malecki, Edward J. "The Economic Geography of the Internet's Infrastructure." *Economic Geography* 78, no. 4 (2002): 399–424.

Malecki, Edward J., and Hu Wei. "A Wired World: The Evolving Geography of Submarine Cables and the Shift to Asia." *Annals of the Association of American Geographers* 99 (2009): 360–82.

Maluc. "The Unregenerate Days." *Zodiac* 21 (August 1928–July 1929): 332.

Marcus, George E. "Ethnography in/of the World System: The Emergence of Multi-Sited Ethnography." *Annual Review of Anthropology* 24 (1995): 95–117.

Mariano, John J., and James C. Baroni. "Enabling Fiber Optic Undersea Cable Technologies for Enhanced Offshore Petroleum Industry Telecommunication Applications." In *Proceedings of the 1997 SubOptic Conference*, 280–87. San Francisco: SubOptic, 1997.

Marshall, K. D. "Cables Can Help Scientists Forecast Lethal Tidal Waves." *Zodiac* 52 (October 1961): 14–15.

Martin, Dan. "Tyco's $75M Maili Facility Going for $5M." *Honolulu Star-Bulletin Business*, April 8, 2004.

Mattern, Shannon. "Infrastructural Tourism." *Places*, July 2013.

McDonald, Philip Bayaud. *A Saga of the Seas: The Story of Cyrus W. Field and the Laying of the First Atlantic Cable*. New York: Wilson-Erickson, 1937.

Miller, Margaret. *Gentlemen of the Cable Service: A Pictorial History of Australia's Overseas Cable Telecommunications Service, 1870–1934*. Sydney: Overseas Telecommunications Commission of Australia and Publimedia, 1992.

Moin, Parvis, and John Kim. "Tackling Turbulence with Supercomputers." *Scientific American* 276, no. 1 (1997): 62–68.

Moore, D. J. "Houng Lee and Waiyamani Win Gold Medals at Pacific Games." *Zodiac* 55 (July 1963): 11.

Morley, David. "Communications and Transport: The Mobility of Information, People and Commodities." *Media, Culture & Society* 33, no. 5 (July 2011): 743–59.

Moss, J. F. "SE, JSN, and HK Will Be First Seacom Link: Retriever (with Fiji Crew) Has Laid Suva Shore Ends." *Zodiac* 53 (September 1962): 5–6.

Müeller-Pohl, Simone. "Colonialism, Decolonization and the Global Media System: The Pacific Telegraph Projects between the United States, Canada and the Cable Companies, 1870–1904." In *Provincializing the United States: Colonialism, Decolonization, and (Post)Colonial Governance in Transnational Perspective*, edited by Eva Bischoff, Norbert Finzsch, and Ursula Lehmkuhl. Heidelberg, Germany: Universitätsverlag Winter, 2014.

Munier, Rob, and Kurt Haaland. "BP GoM: Next Generation Offshore Fiber." *ON&T* 14, no. 7 (2008): 44–45.

Munster, Anna. *An Aesthesia of Networks: Conjunctive Experience in Art and Technology*. Cambridge, MA: MIT Press, 2013.

Nalbach, Alex. "'The Software of Empire': Telegraphic News Agencies and Imperial Publicity, 1865–1914." In *Imperial Co-Histories: National Identities and the British and Colonial Press*, edited by Julie F. Codell, 68–94. Cranbury, NJ: Associated University Presses, 2003.

"The Natives Are Unfriendly (Sometimes)." *Underseas Cable World* 1, no. 3 (1967).

"Neptune." *Underseas Cable World* 1, no. 2 (1967).

"Nerves of the Navy." *Zodiac* 13 (August 1920–July 1921): 270.

"A New Zealand Paper on Cable Improvements." *Zodiac* 15 (August 1923–July 1924): 128.

New Zealand Parliament. "New Zealand Parliamentary Debate: Submarine Cables and Pipelines Protection Bill: Introduction." October 11, 1995. Accessed June 6, 2012. http://www.vdig.net/hansard/archive.jsp?y=1995&m=10&d=11&o=49&p=52.

"No Chinamen." *Zodiac* 17 (January–July 1926): 288.

"Norfolk Island, Australia." *Zodiac* 5 (September 1911): 115–16.

North Star Resources. "Southern Cross Submarine Fiber Optic Cable Landing Project." Accessed April 27, 2012. http://www.nsrnet.com/SouthernCrossMain.htm.

"Of Cables and Conspiracies." *Economist*, February 9, 2008. Accessed July 13, 2014. http://www.economist.com/node/10653963.

Oldenzeil, Ruth. "Islands: The United States as a Networked Empire." In *Entangled Geographies: Empire and Technopolitics in the Global Cold War*, edited by Gabrielle Hecht, 13–42. Cambridge, MA: MIT Press, 2011.

"'Operation Sea Plow': How to Bury a Submarine Cable." *Zodiac* 60 (April 1968): 16–18.

Organisation for Economic Co-operation and Development. "Building Infrastructure Capacity for Electronic Commerce Leased Line Development and Pricing." Paris: Organisation for Economic Co-operation and Development, 1999.

P., B. "A Short Account of La Perouse." *Zodiac* 7 (June 1913–May 1914): 64–65.

Pacific Cable Act 1901. *Account 1907–1908*. London: Her Majesty's Stationery Office, July 1908.

Pacific Cable Board. "Fanning Island-Suva Cables, 1926" and "Telegraph Cables (Further Papers Relating to) No. 13." Papers Presented to both Houses of the General Assembly by Command of His Excellency, New Zealand, 1914.

Pacific Network. "Honotua Pt. 1." Accessed July 5, 2012. http://www.pacificnetwork.tv/watch/honotua-1.

Pannett, Bryan M. J., and Dave A. Hercus. "The ANZCAN Submarine Cable System." Unpublished report.

Parikka, Jussi. "Introduction: The Materiality of Media and Waste." In *Medianatures: The Materiality of Information Technology and Electronic Waste*, edited by Jussi Parikka. N.p.: Open Humanities, 2011.

———. "Mapping Noise: Techniques and Tactics of Irregularities, Interception and Disturbance." In *Media Archaeology: Approaches, Applications, and Implications*, edited by Erkki Huhtamo and Jussi Parikka, 245–77. Berkeley: University of California Press, 2011.

Parkinson, Cecilia. "In England I Was Made to Feel at Home." *Zodiac* 53 (September 1962): 25.

Parks, Lisa. "Around the Antenna Tree: The Politics of Infrastructural Visibility." *Flow* 9, no. 8 (2009).

———. *Cultures in Orbit: Satellites and the Televisual*. Durham, NC: Duke University Press, 2005.

———. "Technostruggles and the Satellite Dish: A Populist Approach to Infrastructure." In *Cultural Technologies: The Shaping of Culture in Media and Society*, edited by Göran Bolin, 67–84. New York: Routledge, 2013.

Parliament of the Commonwealth of Australia, House of Representatives. "Telecommunications and Other Legislation Amendment (Protection of Submarine Cables and Other Measures) Bill 2005: Explanatory Memorandum." 2004–5. Accessed June 6, 2012. http://www.acma.gov.au/webwr/_assets/main/lib100668/expl%20memo%20submarine%20cable%20bill%20pdf.pdf.

Pepperell, Susan. "Hunt for New Species Begins in Titahi Bay." Stuff.co.nz, May 2, 2011. Accessed April 9, 2012. http://www.stuff.co.nz/national/4622175/Hunt-for-new-species-begins-in-Titahi-Bay.

Péron, Françoise. "The Contemporary Lure of the Island." *Tijdschrift voor Economische en Sociale Geografie* 95, no. 3 (2004): 326–39.

Person, Charles W. "Life on Midway Island." *Zodiac* 9 (June 1916–June 1917): 258–59.

Picker, John M. "Atlantic Cable." *Victorian Review* 34, no. 1 (2008): 34–38.

PIPE Networks. "Transmission Technology (Part 2)." PPC-1 Progress Blog. Accessed July 13, 2014. http://www.pipenetworks.com/ppc1blog/category/technology/page/5/.

Polmar, Norman. *The Naval Institute Guide to the Ships and Aircraft of the U.S. Fleet*. Annapolis: U.S. Naval Institute, 2005.

Poltock, F. P. "Bolshevism in Cocos." *Zodiac* 12 (August 1919–July 1920): 152–53.

Porthcurno Telegraph Museum. "Nerve Center of Empire." Accessed July 12, 2014. http://www.porthcurno.org.uk/nerve-centre.

Postman, Neil. "The Reformed English Curriculum." In *High School 1980: The Shape of the Future in American Secondary Education*, edited by Alvin C. Eurich, 160–68. New York: Pitman, 1970.

Pratt, John. "Another Country: Planning Route for ANZCAN." *Zodiac* 27 (1981): 10–13.
Press, Daniel. *Saving Open Space: The Politics of Local Preservation in California*. Berkeley: University of California Press, 2002.
Press, G. W., and B. J. Stringfellow. "The Second Trans-Tasman Cable System." *New Zealand Engineering* 29, no. 12 (December 1974): 333–39.
"Previewing a New Line of Underwater Television Equipment." *Underseas Cable World* 2, no. 6 (1969).
Rauscher, Karl Frederick. *ROGUCCI Study Final Report*. New York: IEEE Communications Society, 2010.
Reinaudo, Christian. "New Market Environment Calls for New Business Practices." In *Proceedings of the 1997 SubOptic Conference*, 13–16. San Francisco: SubOptic, 1997.
"River under Sea Bed." *Zodiac* 27 (August 1934–July 1935): 335.
"Roebuck Bay, Broome." *Zodiac* 4 (April 1910–11): 36.
"Rough Seas, Low Tides Expose Historic Cable." *News-Mail Bundaberg*, April 27, 1990.
Rowe, Ted. *Connecting the Continents: Heart's Content and the Atlantic Cable*. St. John's, NL: Creative, 2009.
Royle, Stephen. *A Geography of Islands: Small Island Insularity*. London: Routledge, 2001.
Rozwadowski, Helen M. *Fathoming the Ocean: The Discovery and Exploration of the Deep Sea*. Cambridge, MA: Harvard University Press, 2005.
Ruddy, Michael. "An Overview of International Submarine Cable Markets." Executive Telecom Briefings, Boston University. December 12, 2006. Accessed July 13, 2014. http://www.terabitconsulting.com/downloads/an-overview-of-international-submarine-cable-markets.pdf.
Russell, Florence Kimball. *A Woman's Journey through the Philippines on a Cable Ship That Linked Together the Strange Lands Seen en Route*. Boston: L. C. Page, 1907.
Russell, Sir William Howard. *The Atlantic Telegraph*. 1865. Stroud, UK: Nonsuch, 2005.
Sampson, Tony. "The Accidental Topology of Digital Culture: How the Network Becomes Viral." *Transformations* 14 (March 2007).
Sandvig, Christian. "The Internet as Infrastructure." In *The Oxford Handbook of Internet Studies*, edited by William H. Dutton, 86–108. Oxford: Oxford University Press, 2013.
Sassen, Saskia. *The Global City: New York, London, Tokyo*. Princeton, NJ: Princeton University Press, 2001.
Schmidt, Michael S., Keith Bradsher, and Christine Hauser. "U.S. Panel Cites Risks in Chinese Equipment." *New York Times*, October 8, 2012.
"Schools 'Adopt' the Company's Cableships." *Zodiac* (November 1961): 18.
Schwoch, James. *Global TV: New Media and the Cold War, 1946–69*. Urbana: University of Illinois Press, 2009.
"Scientists Will Visit Fanning." *Zodiac* 48 (March 1957): 12.
Scott, David. "Dive, Dive and I'll Connect You to Siberia." *New Scientist* 148, no. 2003 (1995): 22.
Scott, R. Bruce. *Gentlemen on Imperial Service: A Story of the Trans-Pacific Telecommunications Cable Told in Their Own Words by Those Who Served*. Victoria, BC: Sono Nis, 1994.

———. "A Short History of the Barkley Sound, in Bamfield Survey, Report No. 1." *Bamfield Marine Station*, April 1970, 52.

"SEACOM Sta. Is Dedicated." *Pacific Journal*, March 31, 1967.

Shackleton, Ernest. "Fanning Island." *Zodiac* 17 (August 1925): 67.

Shapiro, Seymour, et al. "Threats to Submarine Cables." In *Proceedings of the 1997 SubOptic Conference*, 742–49. San Francisco: SubOptic, 1997.

Shimura, Seiichi, ed. *International Submarine Cable Systems*. Tokyo: KDD Engineering and Consulting, Inc., 1984.

"Ships' Anchors Interrupt Shore-Ends of Cables at Penang and Cause Landing Place to Be Shifted Six Miles." *Zodiac* 15 (August 1923–July 1924): 95.

Silverstein, Harvey. "CAESAR, SOSUS, and Submarines: Economic and Institutional Implications of ASW Technologies." OCEANS '78 (September 1978): 406–10.

Simpson, G. S. "*Electra* Looks for Marine Life." *Zodiac* 45 (December 1954): 10.

"Singapore Has Trebled Radio and Doubled Cable Capacity." *Zodiac* 48 (February 1957): 5–6.

Singtel, Ryan. "Cable Cut Fever Grips the Web." *Wired*, February 6, 2008.

Skipper, Mervyn G. "Children of the Submarine Cable Service." *Zodiac* 11 (May 1919): 12.

———. "Shore=Ends: Real and Imaginary." *Zodiac* 11 (May 1919): 22.

"Small Islands." *Zodiac* 11 (April 1919): 225–27.

Smith, Willoughby. *The Rise and Extension of Submarine Telegraphy*. London: J. S. Virtue, 1891.

Smithies, James. "The Trans-Tasman Cable, the Australasian Bridgehead and Imperial History." *History Compass* 6, no. 3 (2008): 691–711.

Southern Cross Cable Network. "About Southern Cross." Accessed July 13, 2014. http://www.southerncrosscables.com/home/company/company.

Stachurski, Richard. *Finding North America: Longitude by Wire*. Columbia: University of South Carolina Press, 2009.

Standage, Tom. *The Victorian Internet: The Remarkable Story of the Telegraph and the Nineteenth Century's On-Line Pioneers*. New York: Walker, 1998.

Star, Susan Leigh. "The Ethnography of Infrastructure." *American Behavioral Scientist* 43, no. 3 (1999): 377–91.

Starosielski, Nicole, Braxton Soderman, and cris cheek. "Introduction: Network Archaeology." *amodern* 2 (2013).

Steinberg, Philip. *The Social Construction of the Ocean*. Cambridge: Cambridge University Press, 2002.

Stephenson, Neal. "Mother Earth, Mother Board." *Wired*, December 1996.

Sterne, Jonathan. *The Audible Past: Cultural Origins of Sound Reproduction*. Durham, NC: Duke University Press, 2003.

Stratford, Elaine, et al. "Envisioning the Archipelago." *Island Studies Journal* 6, no. 2 (2011): 113–30.

Taaffe, Edward J., and Howard Gautier. *Geography of Transportation*. Englewood Cliffs, NJ: Prentice-Hall, 1973.

Tagare, Sunil. "Risk Management Is Key to Submarine Cable Success." Sunil "Neil" Tagare's Personal Views on the Telecom Industry. Accessed April 3, 2012. http://tagare.blogspot.com/2012/03/curse-of-eig-cable.html.

Tarrant, Donald R. *Atlantic Sentinel: Newfoundland's Role in Transatlantic Cable Communications*. St John's, NL: Flanker, 1999.

Tarver, F. H. C. "Forty One Years Foreign Service." *Zodiac* 29 (August 1936–July 1937): 106–10.

Taylor, Herbert. "The First Cable Reached the Island in July 1903." *Guam Recorder*, July 1936.

Taylor, Mark. *Confidence Games: Money and Markets in a World without Redemption*. Chicago: University of Chicago Press, 2004.

TE SubCom. "Micronesian Telecommunications Providers FSMTC and MINTA Contract with Tyco Telecommunications to Construct Undersea Fiber Optic Connections." Accessed May 22, 2012. http://www.subcom.com/company/view.asp?id=313.

"Teleglobe Raises Building Figures." *Alberni Valley Times*, April 8, 1982.

"Terrorist Attack Blasts Haifa Office." *Zodiac* 39 (November 1947): 11.

"Theatricals in the Cocos Islands." *Zodiac* 2 (April 1907–May 1908): 64–66.

"30 Years on, TV's Unresolved." *Review*, Suva, Fiji, August 1993: 11.

"Threatened? Logging Proceeding on MB & PR Timber Property." *Alberni Valley Times*, November 24, 1963.

"TOA Flying Boat Passengers in Fiji." *Pacific Islands Monthly*, June 1948, 26.

Torrieri, Marisa. "Kodiak-Kenai Cable Company Kicks Off Undersea Arctic Fiber Optic Project." *Technology Marketing Corporation-net Mobility Techzone*, January 14, 2010. Accessed July 13, 2014. http://www.mobilitytechzone.com/broadband-stimulus/topics/broadband-stimulus/articles/72684-kodiak-kenai-cable-company-kicks-off-undersea-arctic.htm.

Totton, A. Knyvett. "Notes on Marine Animals and Their Preservation." *Zodiac* 15 (August 1923–July 1924): 314.

Toyn, Stanley L. "Globigerinella Aquilateralis—From the Deserts of the Deep." *Zodiac* 43 (January 1952): 8–9.

Trebett, Meg. "Story of 60-Odd Years of the Bamfield Cable Station." *Alberni Valley Times*, April 5, 1976.

Tsing, Anna. *Friction: An Ethnography of Global Connection*. Princeton, NJ: Princeton University Press, 2005.

Turbin, P. A. "C. S. Recorder to Carry Out Mediterranean Survey." *Zodiac* 60 (August 1968): 15.

"Twins at Fanning Island Are Given Silver Cups from the Chairman." *Zodiac* 51 (May 1960): 7.

"Undersea Seastation." *Underseas Cable World* 1, no. 3 (1967).

United Kingdom Inter-Departmental Committee on Injuries to Submarine Cables. *Report of Inter-Departmental Committee on Injuries to Submarine Cables*. London: Eyre and Spotswoode, 1908.

U.S. Senate. *A Bill to Prevent the Unauthorized Landing of Submarine Cables in the United*

States, Hearings before the Committee on Interstate Commerce, United States Senate, Sixty-Sixth Congress, Third Session, on S. 4301. Washington: Government Printing Office, 1921.

Vaidhyanathan, Siva. *The Googlization of Everything (and Why We Should Worry)*. Berkeley: University of California Press, 2011.

Van Aken, Hendrik M. "Dutch Oceanographic Research in Indonesia in Colonial Times." *Oceanography* 18, no. 4 (2005): 30–41.

Vanderbilt, Tom. *Survival City: Adventures among the Ruins of Atomic America*. New York: Princeton Architectural, 2002.

"Via the Microphone: A Further Talk between 'The Man in the Street' and Mr. H. J. Groom." *Zodiac* 33 (January 1941): 135–37.

Vietnam News Briefs. "Vietnam Asks Telcos to Lay More Telecom Lines after Cable Theft." Vietnam Ministry of Posts and Telecommunications, June 5, 2007.

"A Volcano in the Philippines." *Zodiac* 5 (August 1911): 94–106.

Wagner, Eric. "Submarine Cables and Protections Provided by the Law of the Sea." *Marine Policy* 19, no. 2 (1995): 127–36.

Waianae Neighborhood Board No. 24. "Minutes of Regular Meeting, February 7, 2006." Honolulu: Neighborhood Commission, 2006. Accessed July 13, 2014. http://www1.honolulu.gov/refs/nco/nb24/06/24_2006_02+min.pdf.

———. "Minutes of Regular Meeting, November 6, 2007." Honolulu: Neighborhood Commission, 2007. Accessed July 13, 2014. http://www1.honolulu.gov/refs/nco/nb24/07/24200711min.pdf.

Warf, Barney. "International Competition between Satellite and Fiber Optic Carriers: A Geographic Perspective." *Professional Geographer* 58, no. 1 (2006): 1–11.

———. "Reach Out and Touch Someone: AT&T's Global Operations in the 1990s." *Professional Geographer* 50, no. 2 (1998): 255–67.

Watts, Robert B. "Maritime Critical Infrastructure Protection: Multi-Agency Command and Control in an Asymmetric Environment." *Homeland Security Affairs* 1, no. 2 (August 2005). Accessed July 14, 2014. http://www.hsaj.org/?article=1.2.3.

White, Brian. "Interview with R. Bruce Scott on December 21, 1973, Victoria B.C." In *Selected Oral History Transcripts on the History of Bamfield and Nearby Areas*. Bamfield, BC, Canada: Bamfield Marine Station, 1973.

White House. NSTAC Chair Letter to President Barack Obama and *NSTAC Response to the Sixty-Day Cyber Study Group*, March 2009, 10–11. Accessed May 9, 2012. http://www.whitehouse.gov/files/documents/cyber/NSTAC%20Response%20to%20the%20Sixty-Day%20Cyber%20Study%20Group%203-12-09.pdf.

Whitman, Edward C. "SOSUS: The 'Secret Weapon' of Undersea Surveillance." *Undersea Warfare* 7, no. 2 (Winter 2005). Accessed July 13, 2014. http://www.navy.mil/navydata/cno/n87/usw/issue_25/sosus.htm.

Wilson, George Grafton. "Landing and Operation of Submarine Cables in the United States." *American Journal of International Law* 16, no. 1 (1922): 68–70.

Winseck, Dwayne R., and Robert M. Pike. *Communication and Empire: Media, Markets, and Globalization, 1860–1930*. Durham, NC: Duke University Press, 2007.

———. "The Global Media and the Empire of Liberal Internationalism, circa 1910–30." *Media History* 15, no. 1 (2009): 31–54.

Wood, Matthew P., and Lionel Carter. "Whale Entanglements with Submarine Communications Cables." IEEE *Journal of Oceanic Engineering* 33, no. 4 (2008): 445–50.

Worzyk, Thomas. *Submarine Power Cables: Design, Installation, Repair, Environmental Aspects*. Berlin: Springer-Verlag, 2009.

You, Yuzhu, and Bruce Howe. "Turning Submarine Telecommunications Cables into a Real-Time Multi-Purpose Global Climate Change Monitoring Network." In *Proceedings of* SENSORCOMM *2011: The Fifth International Conference on Sensor Technologies and Applications* (2011): 184–88.

Zahra, Charlot. "More Internet Cable Incidents Likely to Happen in Mediterranean." *Business Today*, January 14, 2009.

"The 'Zodiac' as Geography Primer for Schools." *Zodiac* 28 (August 1935–July 1936): 277.

INDEX

Note: Italic page numbers indicate figures.